# AP Statistics

# *Math Tutor Lesson Plan Series*

# Book 1

iGlobal Educational Services

# iGlobal

## Educational Services

### Believe.Inspire.Transform.

To order, contact iGlobal Educational Services, PO Box 94224, Phoenix, AZ 85070

Website: www.iglobaleducation.com

Fax: 512-233-5389

# AP Statistics

## Contents

# Introduction

Tutoring is beginning to get the respect and recognition it deserves. More and more learners require individualized or small group instruction whether it is in the classroom setting or in a private tutoring setting either face-to-face or online.

This lesson plan book is the fifth out of sixth titles in the series "*Math Tutor Lesson Plan*" Series. It is conceived and created for tutors and educators who desire to provide effective tutoring either in person or online in any educational setting, including the classroom.

## Inside This Lesson Plan Book

This *AP Statistics: Math Tutor Lesson Plan Series* book provides appropriate practice during tutoring sessions for learners for both face-to-face and online tutoring sessions focused on topics in AP Statistics.

The goal of the *AP Statistics: Math Tutor Lesson Plan Series* book is to support all types of tutors. Also, this book is to support teachers who want to provide in-class tutoring to their students in either an individualized or small group tutoring setting. Lastly, this book is also for teachers who are providing math intervention either individually or in small group tutoring sessions either face to face or online so that they can select the specific lesson plan to address the learner's math learning needs.

## How to Use This Lesson Plan Book

iGlobal Educational Services, in collaboration with, Dr. Alicia Holland-Johnson, Tutor Expert and Consultant, created this tutoring resource to help with designing effective tutoring instruction for tutors and teachers who desire to provide in-class tutoring sessions.

These specific lessons were selected based upon field-tested experiences with learners who had learning needs over the years in these specific areas in mathematics. We have provided learning objectives and specific topics covered in each tutoring session so that you can align them with your state's specific standards or adapted standards. For overseas tutors, you can follow suite and align the lesson objectives to specific educational standards required in your country.

These lesson plans should be used to supplement a strong and viable curriculum that encourages differentiation for all diverse learners. They can be used in individual or small group tutoring sessions conducted face-to-face or online in any educational setting, including the classroom.

## Organization of the Lesson Plan Book

Rather than provide a specific "curriculum" to follow, *AP Statistics: Math Tutor Lesson Plan Series* book provides a blueprint to design effective tutoring lessons that are aligned with the "*Dr. Holland-Johnson's Session Review Framework*". Tutor evaluators and coaches are able to analyze tutoring sessions and coach tutors when utilizing the *"Dr. Holland-Johnson's Lesson Plan Blueprint for Tutors"*. In each lesson plan, learners have an opportunity to focus on real-world connections, vocabulary, and practice the math concepts learned in the tutoring sessions in the appropriate amounts to learn and retain the content knowledge. Tutors will have an opportunity to provide direct and guided instruction, while learners practice concepts on their own during independent instruction.

Each lesson plan comes with a mini-assessment pertaining to the math concepts learned in the specific tutoring session. Depending on the learner's academic needs, the tutor or teacher will deem when it is appropriate to administer the mini-assessment. For online tutoring sessions or as an online option to take the mini-assessment, tutors and teachers can upload these mini-assessments to be completed online in their choice of an online assessment tool.

# Lesson 1
# Experimental Design

### Lesson Description

This lesson is designed to help students understand the basic principles of experimental design and analyze the effectiveness of other experiments by breaking down the different components of experiments. Please be sure to utilize the questions to help spark student engagement and cover the vocabulary that is associated with this specific tutoring session. For your own knowledge, sample responses have been provided to guide you as well.

### Learning Objectives

In today's session, the learner will distinguish common terms and basic principles of experimental design. Additionally, the learner will analyze the effectiveness of other experiments by breaking down the different components of experiments with a 75% accuracy or above in 3 out of 4 trials.

### Connect Learning Objectives to Students' Lives

**A.**   Knowing the terms will help students understandthe examples on books, white papers, and otherresource materials.

**B.**   Understanding the basic principles of anexperiment will help students be able to come up with an effective experiment on their own.

**C.**   Students will be able to analyze the effectiveness of other experiments if they are able to break down the different components.

## Introduction

What is your definition of an experiment? Have you conducted an experiment before? What is your understanding of an experiment?

An experiment is a method of studying the effects to a variable of certain experimental conditions. The subjects involved in an experiment are also known as experimental units (humans, plants, animals, etc.).

Factors are the explanatory variables in an experiment; there can be one or more factors in an experiment. Factors may also have different levels each.

In an experiment, experimental units are subjected to conditions or interventions known as treatments.

The different levels of a factor, or the combination of levels or several factors, comprise the treatments in an experiment.

# Direct Instruction: Modeling For You

To have a better grasp of these terms, let's look at an example:

*A farmer would like to study the effect of certain fertilizers and the different amount of sunlight on the growth of crops.*

*The experiment is to be conducted in his farm land using 10 bags of corn kernels, each having the same volume and weight. The farmer will use three types of fertilizers and high and low levels of sunlight.*

Let's take some time to reflect upon this experiment.

In this experiment, answer the following questions:

## Teacher Questions

1. What is the experimental unit?
2. What are the factors?
3. What are the levels?
4. How many treatments are applied?

## Sample Student Responses

1. Corn seeds
2. Type of fertilizers and the amount of sunlight.
3. Three types of fertilizers and two levels of sunlight.
4. Six treatments in total.

Before proceeding to the next part of this lesson, reflect on this question:

**What do you think is the difference between an experiment and an observational study?**

An observational study is merely taking a sample from a population and observing their existing characteristics and their effects on the study being conducted.

Unlike an experiment, an observational study does not have treatment or any intervention applied to the subjects.

## Principles of an Experimental Design

There are three main principles of an experimental design:

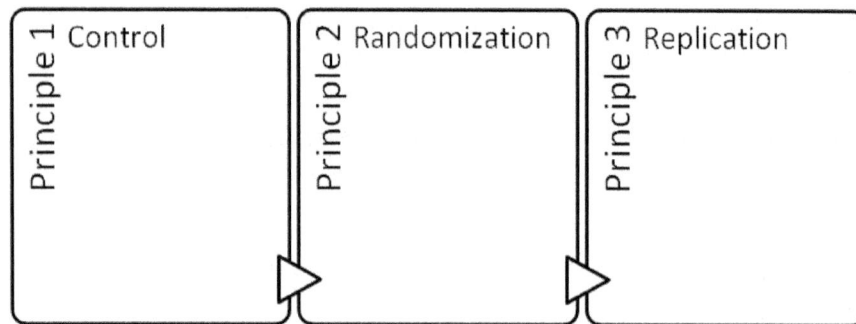

| Principle 1 Control | Principle 2 Randomization | Principle 3 Replication |
| --- | --- | --- |

# Specific Vocabulary for Tutoring Session

### Experiment

An experiment is a method of studying the effects to avariable of certain experimental conditions. The subjects involved in an experiment are also known as experimental units (humans, plants, animals, etc.).

### Factors

The explanatory variables in an experiment;there can be one or more factors in an experiment. Factors may also have different levels each.

### Experimental Units

In an experiment, experimental units are subjected to conditions or interventions known as treatments.

### Levels

The different levels of a factor, or the combination of levels or several factors, comprise the treatments in an experiment.

### Treatment

A term used to describe conditions or interventions in an experiment.

## Control Group

Two or more treatments in an experiment should be comparable enough by employing control variables or placing the experimental units in a controlled environment to avoid effects of extraneous factors.

## Randomization

Experimental units should be assigned to treatments in a random, yet systematic, manner to ensure that there is no bias or favoritism.

## Replication

The experiment should be done in as many subjects as possible; having only one or two subjects will not produce reliable results and will just merely be anecdotes.

## Placebo

This is a kind of treatment that will produce no effect, in order to have more control to the variables. An example would be applying a known treatment (placebo) to a certain group, compared to applying a new test drug to another group.

## Placebo Effect

A phenomenon where the subjects (humans) administered the placebo treatment have shown responses.

## Blinding

A characteristic of an experiment wherein those who are associated with it have no idea how the subjects are assigned into the treatment groups.

## Block Design

An experimental units are grouped, or blocked, a certain way to add more control into the experimentand see clearer results of each treatment applied. An example of a block design in an experiment that was conducted on 500 individuals, thesubjects are blocked by gender – 250 males and 250 females.

## Double-Blinding

A much ideal characteristic of an experiment wherein both the subjects and the evaluators (e.g. physicians, judges) are "blinded".

## Experimental Study

A type of study that involves a treatment, procedure, or program is intentionally introduced, tested, and analyzed for results to be shared with a specific targeted audience.

## Observational Study

A type of study that merely takes a sample from a population and observing their existing characteristics and their effects on the study being conducted. Unlike an experiment, an observational study does not have treatment or any intervention applied to the subjects.

# Guided Instruction: Working With You

**Experiment\*:**

*A biologist is interested in studying the effect of growth-enhancing nutrients and different salinity (salt) levels in water on the growth of shrimp. The biologist has ordered a large shipment of young rock shrimps from a supply house to use for the study. The experiment is to be conducted in a lab where there are 12 tanks (with equal amount of shrimp in each) in a controlled environment.*

*The biologist is planning to use 3 different growth-enhancing nutrients (A,B, and C) and two different salinity levels (high and low).*

## Reflective Questions for this Experiment

1. How many treatments will be done in this experiment?

2. What is the experimental unit?

3. How many factors are applied in this experiment? What are they?

4. What are the different levels of the factor/s?

5. How many treatments are applied?

6. What is an advantage and a disadvantage of having only rock shrimp in the experiment?

7. Is there more control in the experiment with only one kind ofshrimp?

8. How will the result of the experiment be different when different kinds of shrimp are used?

# Video Suggestions for Tutoring Sessions

Please conduct a search on either YouTube or Teacher Tube to find appropriate videos for this lesson. Below are some suggested title searches:

1. Experimental Designs

2. Observational Studies

3. Placebos and Treatments in a Research Study

# Independent Instruction: Working On Your Own

## Questions

☞ <u>**Experiment 1:**</u>

*A farmer would like to test the effect of fertilizers and the amount of sunlight on the growth of crops. 4 bags of rice grains and 4 bags of corn seeds (each with the same volume and weight) will be used in this experiment and will be done in a controlled environment. The farmer plans to use three new fertilizers in the market and high and low levels of sunlight.*

**Reflective Questions for this Experiment**

1. How many blocks and treatments will be done?

2. What is an advantage and a disadvantage of using rice and corn in this experiment?

**Explanation**

Since there two types of crops used, there will be 4 blocks –rice block and corn block. For each crop, there are 6 treatments applied – 3 fertilizers by 2 levels of sunlight. Hence, there are a total of 24 treatments that will be done in this experiment. Using 2 crops will possibly provide more generalized results in terms of the effects of fertilizers and sunlight on the growth of crops. However, this will also bring about possibility of having other factors affect the results such as the difference in nature of the crops, the different requirements of each crop to hasten its growth, etc.

## Mini-Assessment

☞ **Experiment 1:**

*A study is to be conducted to identify the effects of eating sweets in relieving the stress of individuals at their workplace. A group of 160 young professionals will be participating in this study. At any time, they are able to withdraw from the study.*

**Reflective Question for this Experiment**

**1.** How would you propose a design on how the subjects can be grouped?

☞ **Experiment 2:**

*A pharmaceutical company wants to compare the effects of their newly released drug in the market which helps relieve chronic back and neck pain. 1500 patients who have been repeatedly complaining of chronic back and neck pain are the subjects of this study. After 4 weeks of continuously taking the drug, the patients are asked for any improvement.*

**Reflective Questions for this Experiment**

**1.** What is the experimental unit and factor/s in this experiment?

**2.** What are the experimental treatments to be applied?

**3.** Is it possible for this experiment to be double-blinded? Why or why not?

# Mini-Assessment Answers and Explanations

1.  Since male and female deal with stress differently, it's best to block the subjects by gender – 80 males and 80 females. 40 males will be given a chocolate bar each, while the other 40 do not get chocolates. Same is done with the female group. Results are noted per block, and overall.

2.  The experimental unit here is the patient of chronic back and neck pain. The only factor involved in this experiment is the new medicine.

    The different treatments to be conducted are as follows:

    750 patients to be treated with the new drug

    750 patients to be treated with another known drug(placebo)

    Yes, it's possible for this experiment to be double-blinded by not informing the patients nor the physicians what drug the patients will be taking.

.

## Lesson Reflection

We have learned the following concepts in this lesson:

➤ Distinguishing common terms and basic principles of experimental design.

➤ Analyzing the effectiveness of other experiments by breaking down the different components of experiments.

# Lesson 2
# Data & Normal Distributions

## Lesson Description

This lesson is designed to help students understand the different types of data that is associated with both data and normal distributions. Please be sure to utilize the questions to help spark student engagement and cover the vocabulary that is associated with this specific tutoring session. For your own knowledge, sample responses have been provided to guide you as well.

## Learning Objectives

In today's session, the learner will analyze different types of data that includes normal distributions with at least 75% or above accuracy in 3 out of 4 trials.

## Connect Learning Objectives to Students' Lives

A.  Understand the difference between different types of data.

B.  Understand ways of representing data.

C.  Know how to basically analyze some data with special features, that is normal distributions.

D.  Highlight some applications of the concepts learned.

## Introduction

Data is a very important information that makes us understand the nature and features of most things in day to day life. It is simply a collection of information about a given feature, thing or people. Data may be used to inform of numerical values or simply grouped into categories. To understand what data means and imply, we need to represent and analyze it. In most cases, most data shows normality when the sample six is sufficiently large. This makes it easy to draw conclusions from such data, however, in some cases when the sample size is small, less than 30, we have to come up with ways of determining if the data is normal or not. In this lesson, we are going to introduce different types of data, represent it and tell if they are normal or not.

# Specific Vocabulary for Tutoring Session

### Categorical data

This is data that is described in terms of qualitative descriptions. This may be in terms of names or categories of items.

### Numerical data

This is data that is described in terms of quantitative descriptions. It uses numerical numbers to describe a particularfeature of data.

### Discrete Probability Distribution

This type of distribution requires afinite number of values and the random variables are whole numbers.

### Continuous Random Variable

A random variable

### Continuous Probability Distribution

This type of distribution involves rational and irrational numbers and this random variable is referred to as a continuous random variable.

### Probability Density Function

The formula that describes the distribution of a continuous random variable.

### Density Curve

A smooth curve when probability density function is drawn on a plane.

# Direct Instruction: Modeling For You

## Numerical and Categorical Data

### Categorical data

This is data that is described in terms of qualitative descriptions. This may be in terms of names or categories of items. For instance, data that given different colors in any given collection, the names of religions among others. The categories do not reflect quantities. Categorical data are measured using two types of scales. These are nominal and ordinal scales.

In nominal scales, the responses do not have some kind of ordering. These may be names or categories of some items. An example of categorical data with nominal scale is data gender representation in a place. The responses here are males and female only. Notice that these categories cannot be ordered in any way, we cannot say that either one of the gender appears in the first category or second.

In ordinal scales, data is grouped in categories that can be ordered. For instance, we may be interested in knowing the color intensity of various pigments used to die clothes. We may say, pigments with highest intensity, mild intensity and lower intensity. In this case, the categories can be ordered based on the intensity and they still represent qualitative characteristics of the items being ordered.

### Numerical data

This is data that is described in terms of quantitative descriptions. It uses numerical numbers to describe a particular feature of data. This feature may be height, weight, length, time among others. There are two scales that describes numerical data, these are interval and ratio scale.

In the interval scale, the responses have an equal interval that represents the same physical quantity. For instance, we may consider time in 12 clock hour system. The difference between 2 o'clock and 4 o'clock is the same as the time difference between 9 o'clock and 11 o'clock is the same; it shows a similar time interval. However, this scale does not have absolute zero. In 12 clock hour system, we do not have 0 hours as in 24 clock hour system.

In ratio scales, all the features of nominal, ordinal and interval scale. In addition, the scale has an absolute zero, which implies absences of the quantity.

### Graphical Displays

This is one of the most common ways of representing data. It shows the geometric implications of a given data. The choice of a given type of graph to use depends on the type of data and the intended use of the data.

Below are some types of graphs:

## Bar graphs

This composed of vertical parallel bars that represent different names or categories. The bars are drawn with gaps between them. The height of a bar is proportional to the number of items in a given category. They are best for representing both categorical and numerical data. They are mostly used to compare things and track changes over a time.

The horizontal axis bears the categories or names and the vertical axis their frequencies.

## Illustration

Draw a bar graph showing the performance of a student in different test shown below:

| Test | Score (%) |
|------|-----------|
| First test | 55% |
| Second test | 50% |
| Third test | 60% |
| Fourth test | 75% |

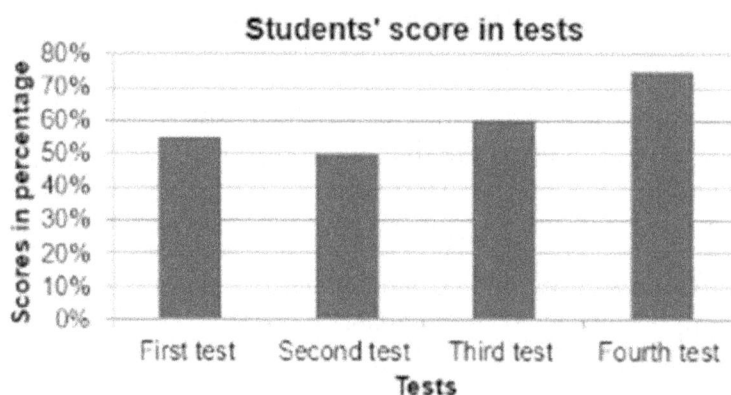

## Line Graph

A line graph is a graph where the frequencies of the different categories are connected to come up with a line. They are used to compare items and track changes among the given categories or names. They are better than bar graphs is showing changes over a short time.

Using the data above, the corresponding line graph will be as shown below:

## Students' score in tests

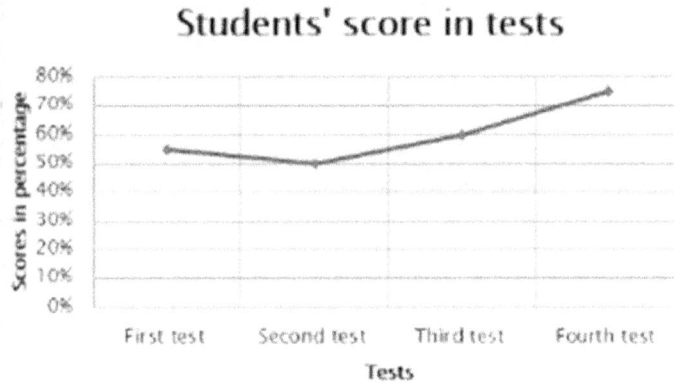

## Histogram

A histogram is a graphical display used to show the distribution of data grouped in ranges. It has parallel bars except that there is no spaces between them. A space may only appear when a given range has no a frequency of zero.

Unlike the bar graph, where the horizontal axis has categories, the histogram has number ranges. For this reason, it is suitable for representing continuous data such as length, weight, mass among others.

## Illustration

The data below shows the weight ( in pounds) of students sampled from grade 5 to grade 7.

60, 64, 65, 78, 63, 87, 69, 73, 74, 75, 72, 76, 83, 78, 75, 80, 79, 67, 70

When the data is put in groups of interval 5, we have the following:

| Ranges | Number of Students |
|--------|--------------------|
| 60-64  | 3                  |
| 65-69  | 3                  |
| 70-74  | 4                  |
| 75-79  | 6                  |
| 80-84  | 2                  |
| 85-89  | 1                  |

At this level, we will go into the details of representing grouped data on the a histogram, because we are only concerned about the differences between it and a bar graph. The values to be used on the horizontal axis are the first lowest value in the first group and all the largest values in every group.

Upon plotting we have the following graph:

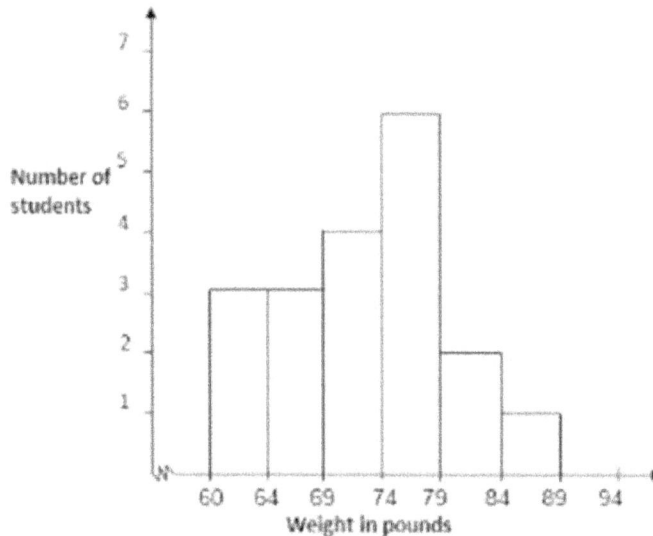

## Pie Chart

A pie chart is a circular graph where it is divided based on the proportions of a category in relation to the total number of items. It is best used when one is interested in the percentage of a given category.

## Illustration

A survey was done to determine the number of people with different eye colors among the employees of a community based organization. Represent the data in a pie chart.

Use the pie chart to determine the percentage of people with green eyes more than those with hazel eyes.

| Eye color | Number of Students |
|-----------|--------------------|
| Brown     | 26                 |
| Blue      | 9                  |
| Green     | 8                  |
| Hazel     | 5                  |
| Amber     | 2                  |

The percentage and the corresponding angle representation are calculated as shown below. The total number of people is $26 + 9 + 8 + 5 + 2 + 50$

The percentages are,   Brown $= \dfrac{26}{50} \times 100\% = 54\%$          Hazel $= \dfrac{5}{50} \times 100\% = 10\%$

Blue $= \dfrac{9}{50} \times 100\% = 18\%$          Amber $= \dfrac{2}{50} \times 100\% = 4\%$

Green $= \dfrac{8}{50} \times 100\% = 16\%$

Since the sum of angles at a point is $360°$, the angles will be

$Brown = \dfrac{26}{50} \times 360 = 187.2°$          $hazel = \dfrac{5}{50} \times 360 = 36°$

$Blue = \dfrac{9}{50} \times 360 = 64.8°$          $Amber = \dfrac{2}{50} \times 360 = 14.4°$

$Green = \dfrac{8}{50} \times 360 = 57.6°$

The angles are then used to divide the pie chart as shown below.

**Employees' eye color**

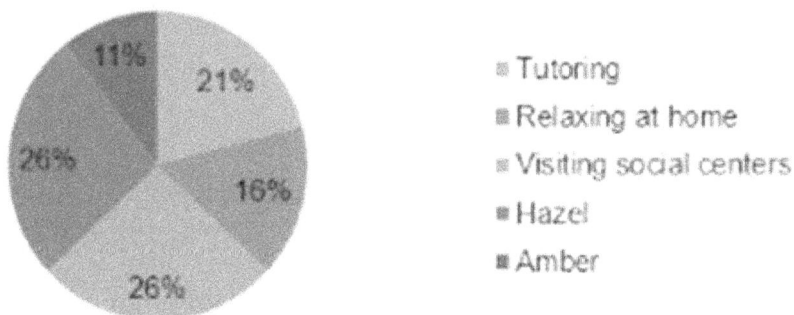

11%  21%  26%  16%  26%

- Tutoring
- Relaxing at home
- Visiting social centers
- Hazel
- Amber

The percentage of those with hazel eyes is 10% while those with green eyes is 16%.

The difference is $16\% - 10\% = 6\%$.

## Box Plot

Box plot is a graph with a horizontally or vertically oriented box whose one ends represent the upper and the lower quartile. When the box is vertical, then there are vertical lines drawn from the end of the box to the lower and highest values of the data, the lines are called whiskers.

The graph is also called the box-and –whiskers plot. A line is drawn within the box to mark the median of the data.

## Illustration

Draw the box-whiskers-plot for the following data:

20, 25, 32, 22, 27, 24, 23, 28, 24

We determine median and the quartiles of the data, we arrange them in ascending order.

20, 22, 23, 24, 24, 25, 27, 28, 32

The median value also called the second quartile is Q2 = 24

The lower quartile is the median of the first half = $Q1 = \frac{23 + 22}{2} = 22.5$

The upper quartile is the median of the second half = $Q3 = \frac{28 + 27}{2} = 27.5$

The lowest value of the data is 20

The highest value of the data is 32

The plot is

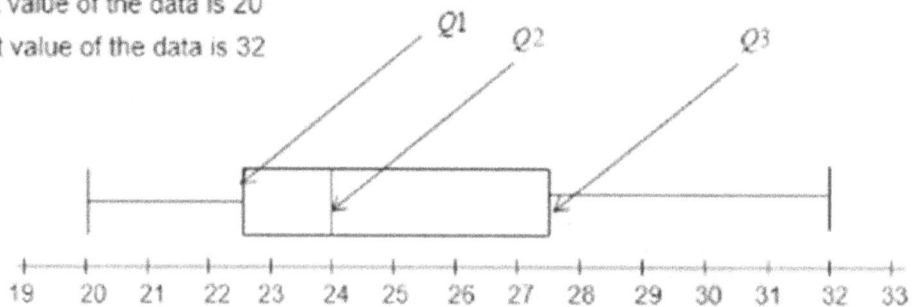

## Numerical Displays

This is where data is represented using numerical values and not graphs. The most commonly used numericaldisplay is the leaf and stem plot.

## Leaf and Stem Plot

This is a display where the numbers split into two parts, the leaf and the stem. The leaf is usually composed of theplace value of the last digit while the stem is the rest of the digits to the left.

## Illustration

Given the data 78, 30, 24, 39, 39, 40, 55, 72, 32

The last figure in any given number is ones, hence the leaf will be composed of the ones while the stem will be tens.

78 will be 7 and 8 and so on. The plot will be the following:

| Stem | Leaf |
|------|------|
| 2 | 4 |
| 3 | 0 2 9 9 |
| 4 | 0 |
| 5 | 5 |
| 7 | 2 8 |

To know the number of elements, you simply count the number in the leaf column, which are 9.

The values in a leaf and stem plotare arranged in ascending order.

The second row represents 30, 32, 39 and 39.

## Normal Distributions

Consider the following distribution of a random variable:

When we toss a fair coin, when we expect a tail or a head to show up. When we toss it twice, there then we expect four different outcomes. When we let a tail to be a random variable, then in two tosses, we can have 0 tail, 1 tailor two tails.

Since there are 4 outcomes,0 tail implies a head and a head.

The probability of a head and a head is $P(H)P(H) = \frac{1}{2} \times \frac{1}{2} = \frac{1}{4} = 0.25$

1 tail implies a head and a tail or a tail and a head.

Thus $P(H)P(T) + P(T)P(H) = \frac{1}{4} + \frac{1}{4} = 0.25 = 0.25 = 0.5$

2 tails implies a tail and a tail. Thus $P(T)P(T) =$

The distribution of theta random variable will be $= \frac{1}{2} \times \frac{1}{2} = \frac{1}{4} = 0.25$

| Random variable | Probability |
|-----------------|-------------|
| 0 | 0.25 |
| 1 | 0.5 |
| 2 | 0.25 |

Notice that the total probability is 1. This distribution is called discrete probability distribution because it takes a finite number of values and the random variables are whole numbers. In this case, it takes 3 values. In some cases, we have a distribution that is given by a formula where the random variable takes both whole numbers and numbers that are not whole numbers; more precisely, rational and irrational number. Such a random variable is referred to as a continuous random variable while the probability distribution of a continuous variable is called acontinuous probability distribution.

The formula that describes the distribution of a continuous random variable is called a probability density function. When probability density function drawn on a plane, it makes a smooth curve whose total area under it represents the total probability. This smooth curve is called the density curve.

Just like the sum of the probability in a discrete probability distribution, the sum of the probabilities which is the area under the density curve is 1.

The main properties of a density curve is that:

**(i).** The highest point of the curve is the mode of the distribution.

**(ii).** The area under the curve is equal to 1.

**(iii).** Between any given range of values on the horizontal axis, the observations that lie within the values are proportional to the area above the range and under the curve.

**(iv).** The mean and the median of the curve lies at the same point, the middle point, if the curve is symmetrical.

Distributions have a number of descriptions which are worthy noting. Based on these distributions, we are able to draw important conclusions from them. In this lesson, we will not go into their details but just mention them.

The most common descriptions used are the following:

**(i).** The measures of central tendencies. These are values that can be used to represent the whole data. They are the mean, the median and the mode.

**(ii).** The measures of spread. These are values that explains how the values of the data are distributed from the mean. They include the variance, the standard deviation, the range and quartiles.

## Normal curves

These refers to density curves that have a dome shape. They are probability distributions for normal distributions.

For easy handing of these curves, they are standardized so that their mean, $\mu = 0$ and standard deviation, $\sigma = 1$.

The standardization of the normal distribution is where the random variable $X$ is transformed into a randomstandard variable $Z$ which has the mean of zero and a standard deviation of 1 where

$$Z = \frac{x - \mu}{\sigma}$$

The most important feature of the standardized normal curve is that it is symmetrical hence it is easy to describe its distribution using the measures of spread. This description is known as the empirical rule.

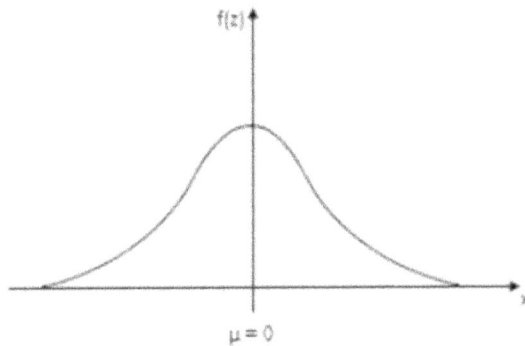

## Empirical Rule and Assessing Normality

The empirical rule explains the distribution of a data within the three standard deviations of a normal distribution.

This rule helps one get estimates of the distribution, after determining the standard deviation, without significantly getting the whole data.

This rule is also called the 68 – 95 – 99.7 rule.

It states that when data has a normal distribution, then a significantly greater percentage of data will lie within the three standard deviations where:

68% of the data will lie within the first standard deviation, $(\mu \pm \sigma)$

95% of the data will lie within the first two standard deviation, $(\mu \pm 2\sigma)$

99.7% of the data will lie within the first three standard deviation, $(\mu \pm 3\sigma)$

We can use this rule to solve some problems related to standard normal distribution.

68% of the data will lie within the first standard deviation, $(\mu \pm \sigma)$

95% of the data will lie within the first two standard deviation, $(\mu \pm 2\sigma)$

99.7% of the data will lie within the first three standard deviation, $(\mu \pm 3\sigma)$

We can use this rule to solve some problems related to standard normal distribution.

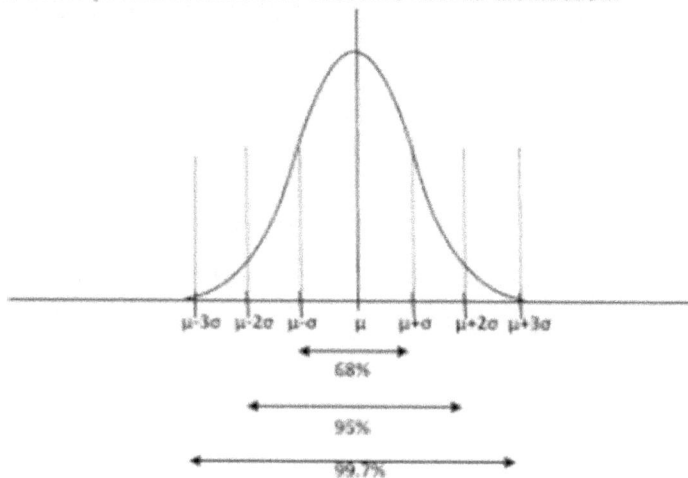

## Assessing Normality

The central limit theorem assures us that when the sample size is more than 30, the data drawn from a normally distributed population. However, when the sample size is less than 30, there should be ways of determining if the data is normally distributed or not. It is important to determine the normality of data for samples less than 30 because most statistical and research experiments uses samples sizes that are less than 30.

To determine if the data is drawn from a population that is normally distributed, we consider the following methods;

**(i).** We use the leaf-and-stem plot or the histogram.

If the shape of the histogram or the leaf-and stem plot is belly shaped or approximately belly shaped, then the data is normally distributed.

**(ii).** The use of the empirical rule

In this method, first determine the mean and the standard deviation of the distribution

We find out if 68% of the data lies between the first standard deviation, 95% lies within the first two standard deviation and 99.7% lies within the first three standard deviations.

**(iii).** The use of the normal probability plot

We first arrange the data values in ascending order then give then positions

For each data value, we find the corresponding percentile using the formula:

$$\frac{\text{position of a data value} + 0.5}{\text{Number of data value}} \times 100$$

We then get the $z$ value associated with the percentile.

When then plot the $z$ values against the data values.

If the plots are linear or approximately linear, then the data is normally distributed.

If the plots are not linear, it depicts a curve, then the data is not normally distributed.

The most commonly used methods are the last two.

# Guided Instruction: Working With You

## Question 1

In which category, would you classify the following data:

**(i).** Clubs in school

**(ii).** Music club

**(iii).** Drama club

**(iv).** Debate club

### Response

The data is based on names of collections of people and does not have any quantitative idea. Therefore, it is categorical data.

## Question 2

How can you manipulate the data on performance of students in an examination to given in terms of percentage to describe categorical data?

## Response

The performance of students given in terms of percentage a numerical data since the responses are given interms of numerical value. To be more precise, it can be described in terms of ratio scale. When the marks are categorized as poor, fair, good, excellent where poor represent 0% to 40%, fair represent 41% to60%, good from 61% to 79% and excellent above 80%, then the data will be categorical. The responses will be poor, fair, good and excellent. The responses have some order, since excellent is ranked the highest and poor the lowest, thus ordinal scale.

## Question 3

List two similarities and two major differences between a bar graph and a histogram,

## Response

Both the bar graph and the histogram are represented using vertical parallel bars. Furthermore, their vertical axes shows the frequency of the categories or number ranges used on the vertical axis. The difference are; the bars of a bar graph have gaps in between them while those of the histogram do not have them. Second, the horizontal axes of a bar graph has categories while that of the histogram has number ranges.

## Question 4

The pie chart below shows the Saturday daytime schedule of a teacher. How long does she spent on tutoring than relaxing at home?

### Teachers day time schedule

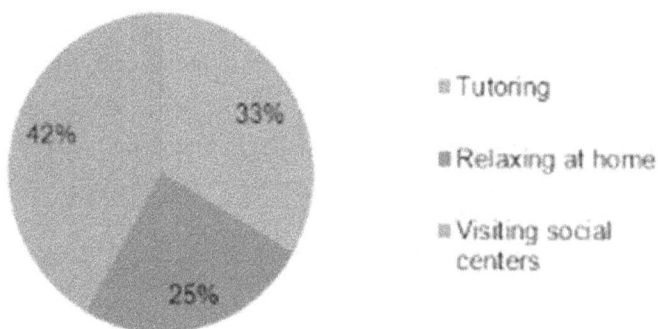

- Tutoring
- Relaxing at home
- Visiting social centers

## Response

The percentage time spent on tutoring is 42%

The percentage time spent relaxing at home is 33%

The difference is 42% - 33% = 9%

Since the whole represent hours in a daytime = 100%, the difference will rerepsent

$\dfrac{9}{100} \times 12 = 1.08 \ hours$

## Question 5

The figure below the results of a survey done to establish the consumption of water by a number of families in ruralareas of USA per month. The consumption was measured in thousands of gallons.

| Stem | Leaf |
|------|------|
| 0 | 4 5 8 9 |
| 1 | 0 1 2 2 3 3 3 4 |
| 2 | 0 1 |
| 3 | 0 |

0 stem and 1 leaf implies 0 1

(i).   How many families were involved in the survey.

(ii).   How many families consume between 700 gallons and 10400 gallons.

(iii).   Find the percentage of the families that consume less than the mean consumption.

## Response

(i).   To know the families that were involved, we count the number of leafs.

   15 families

(ii).   700 gallons = 0.7 thousand gallons

   10400 gallons = 1.04 thousand gallons

   The value that fall between 0.7 and 1.04 are 0.8, 0.9 and 1.0.

   Thus, we have only 3 families.

**(iii).** The mean will be

$$\bar{x} = \frac{0.4 + 0.5 + 0.8 + 0.9 + 1.0 + 1.1 + 2(1.2) + 3(1.3) + 1.4 + 2.0 + 2.1 + 3.0}{15} = \frac{19.5}{15} = 1.3$$

The mean is 1.3 thousand gallons

The values below 1.3 are 0.4, 0.5, 0.8, 0.9, 1.0, 1.1, 1.2, 1.2; 8 values .

The percentage will be $\frac{8}{15} \times 100\% = 53.33\%$

## Question 6

A researcher investigated the number of anopheles mosquito in an area that has had high fatalities due to malaria. She establishes that there is an average of 230 such mosquitoes with standard deviation of 40.

What is the probability that an area sampled within the region of study has the number of anopheles mosquitoes between 150 and 310?

## Response

We list what we have:

The mean, $\mu = 230$ , standard deviation, $\sigma = 40$

The bounds are 150 and 310

We first express the bounds, 150 and 310 in terms of the mean and the standard deviation

$310 = 230 + 80 = 230 + 2(40) = \mu + 2\sigma$

$150 = 230 - 80 = 230 - 2(40) = \mu - 2\sigma$

Thus we have, $\mu \pm 2\sigma$

The bounds lies between the first two standard deviations hence the probability is 95%.

## Question 7

In a mathematics test, 99.7% of the students scored between 46% and 78%. If their marks are normally distributed, what is the mean and the standard deviation of the scores?

## Response

We determine the mean.

Since the distribution is normal, the mean is half way.

The mean $= \dfrac{78 + 46}{2} = \dfrac{124}{2} = 62$

The mean is 62%

We determine the standard deviation.

We first find the number of standard deviations based on the percentage given.

99.7% implies the data lies within the first three standard deviations. Thus, 3 standard deviations on either side implies a total of 6 standard deviations.

Thus, the 46% is 6 standard deviations from 78%.

One standard deviation will be $\sigma = \dfrac{78 - 46}{2} = \dfrac{32}{2} = 16$

Thus the standard deviation is 16%.

## Question 8

Determine if the following data is normally distributed.

53, 78, 40, 56, 81, 41, 61, 69, 64, 53, 55, 69, 68, 54, 79

**Response**

We use the leaf and stem plot where the value in the ones are the leafs and those in the tens positions are the stems.

Thus, we get the following display:

| Stem | Leaf |
|------|------|
| 4 | 0  1 |
| 5 | 3  3  4  5  6 |
| 6 | 1  4  8  9  9 |
| 7 | 8  9 |
| 8 | 1 |

Since the display is approximately normal with the peak between the group having 50's and 60's, the distribution is normal.

# Video Suggestions for Tutoring Sessions

Please conduct a search on either YouTube or Teacher Tube to find appropriate videos for this lesson. Below are some suggested title searches:

1. Normal Distribution in Aquatic Environment

2. The Use of Graphical Displays in Market

3. Application of line graph medicinal research, such as in Diabetes Management Systems to monitor sugar levels and medicine administration

# Independent Instruction: Working on Your Own

(i).   The data below shows the temperature amounts of desert. Draw a line graph to show the trend of temperature amounts in various days and identify two days with the highest rainfall. Explain how you have got them.

| Day | Mon | Tue | Wed | Thu | Fri | Sat | Sun |
|-----|-----|-----|-----|-----|-----|-----|-----|
| Temp. (°C) | 36 | 35.5 | 35 | 37 | 34.5 | 35 | 36.5 |

(ii).   There is 68% chance that most grade 5 students will be between 10.5 and 13 years old. What is the average age of a grade 5 student given that their age is normally distributed.

(iii).   The figure below shows the monthly expenses of Mr. Julius. If he spend $130 on transport, how much in total, does he spent on food and clothing?

## Amounts spent every month

⬛ Food  ⬛ Clothing  ⬛ Transport  ⬛ Savings

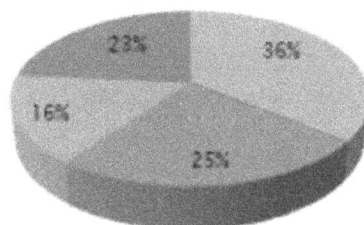

23%  36%  16%  25%

**1.**

### Temperature distribution of a hot desert

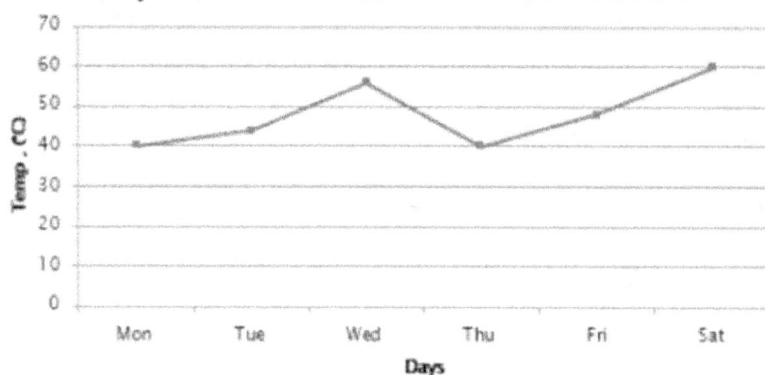

The days with the highest temperature is Thursdays and Sunday

To identify them, we identify two days with the highest peaks.

**2.** Since their age are normally distributed, using the empirical rule, their average will be half-way 10.5 and 13

$$\text{Average age} = \frac{10.5 + 13}{2} = \frac{23.5}{2} = 11.75 \text{ years}$$

**3.** We list what we have

Transport represents 16% and is equivalent to $130

Clothing is 25%

Food is 36%

We then determine the sum of percentage of food and clothing

The sum is 25% + 36% = 61%

We use proportions to determine the amount equivalent to 61%

Let the amount be $x$, then

$$\frac{x}{61} = \frac{130}{16}$$

Cross multiplying we get

$$16x = 130 \times 61$$

$$x = \frac{130 \times 61}{16} = \$496$$

The two expenses cost $496

# Mini-Assessment

1.  Among the following, which one is a numerical data

    A.  Red, violet, green

    B.  Blue = 30, Yellow = 23, Orange = 27

    C.  Grade 2 = 20, grade 3 = 18, grade 4 = 22

    D.  1 – 5-------3
        6 – 10-----2
        11 – 15----6

    E.  Male = 5, female = 6

2.  Identify the best display that would best represent a continuous data showing the trend over time.

    A. Pie chart          B. Leaf - and - stem plot          C. Histogram

    D. Bar graph          E. Box plot

3.  Which feature that makes one differentiate between a histogram and a bar chart?

    A. Title

    B. Contents of the horizontal axis

    C. Contents of the vertical axis

    D. The size of the bars

    E. Orientation of the graph

4.  There is 95% chance that the height of an adult male will fall between 60 inches and 74 inches. If the height is normally distributed, what is the standard deviation of this distribution.

**5.** A survey was done in a town to determine the number of animal pets a family has. The data below shows the results of the survey.

2, 3, 4, 1, 0, 2, 3, 2, 1, 4, 5, 6,4

Draw a box plot to show the distribution of animal pets in the sample

**6.** The mean and the standard deviation of eight subjects Eliud is taking is 68% and 5% respectively. If his marks are normally distributed, what is the probability that he scores between 53% and 83% in a subject.

**7.** The figure below shows the amount of mobile phones sold by Capital mobile shop limited in a week. If two thirds of the phones sold on the previous Friday is half that sold on Saturday, how many phones were sold on previous Friday.

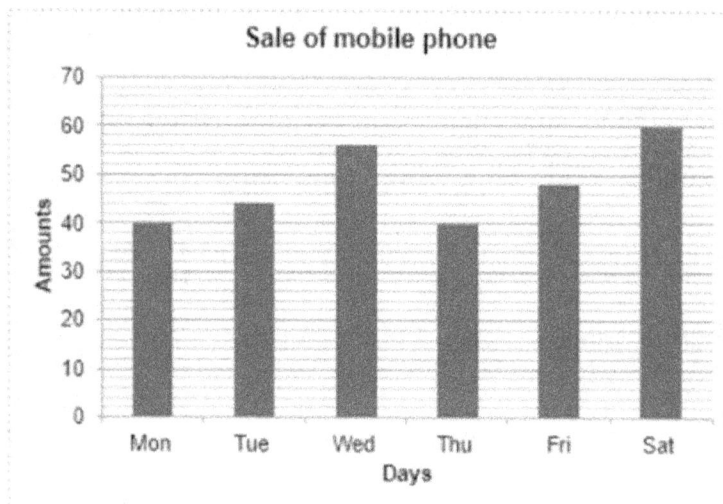

# Mini-Assessment Answers and Explanations

**1.** D

Numerical data is composed of numerical values only. Thus the correct answer is option D.

**2.** C

The best graphs that represents continuous data that shows a trend over a long time is a line graph and a histogram. Among the options given, the correct answer is a histogram.

**3.** B

The difference between a histogram and a bar chart is that the bars of a bar graph have gaps in between them while those of the histogram do not have them. Secondly, the horizontal axes of a bar graph has categories while that of the histogram has number ranges.

Based on the options given, the difference lies on the contents of the horizontal axis.

**4.** 95% chance represents 4 standard deviations, 2 to the left and 2 to the right

The 4 standard deviations are equivalent to the difference $74 - 60 = 14$ inches

The standard deviation $= \dfrac{14}{4} = 3.5$ inches

**5.** We arrange the data in ascending order so as to determine the median and the lower and upper quartiles.

0, 1, 1, 2, 2, 2, 3, 3, 4, 4, 4, 5, 6.

There are 13 values

The least value is 0 while the larges one is 6

The median value is in position 7, that is 3

The median of the first half = third quartile = 4

The median of the school half = third quartile = 4

The plot is

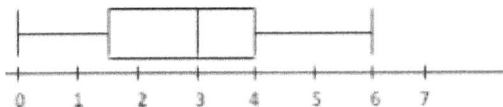

**6.** $\mu = 68 \quad \sigma = 5$

The bounds are 53 and 83

We express the bounds in terms of mean and standard deviation

$53 = 68 - 15 = 68 - 3(5) = \mu - 3\sigma$

$83 = 68 + 15 = 68 + 3(5) = \mu + 3\sigma$

Thus the values lies within 3 standard deviations from the mean. Using empirical rule, the probability is 99.17%

$= 0.997$

**7.** On Saturday, 60 phones were sold

Half of these sold is 30

30 is equivalent to two thirds of those sold previous Friday

Hence the amount of phones sold on previous Friday $= \dfrac{3}{2} \times 30 = 45$ phones.

## Lesson Reflection

In this lesson, we have discussed what categorical and numerical data are. We have gone forward and seen how we can represent these types of data graphically and also numerically, where possible. We highlighted some ways of describing a distribution. We have also looked at density curve and discussed a special density curve, the normal curve and its distribution. Under this sub topic, we have discussed the empirical rule that enables us predict the characteristics of a normal curve without having all the data. We have concluded by looking at the ways of assessing if data is normally distributed.

# Lesson 3
# Linear and Nonlinear Bivariate Data

## Lesson Description

This lesson is designed to help students understand the difference between response and explanatory variables. Additionally, students will have an opportunity to analyze data using both scatter and residual plots. Please be sure to utilize the questions to help spark student engagement and cover the vocabulary that is associated with this specific tutoring session. For your own knowledge, sample responses have been provided to guide you as well.

## Learning Objectives

**(i).** To discuss the difference between response and explanatory variables.

**(ii).** To Represent a bivariate data graphically using scatter plot.

**(iii).** To analyze data using scatter and residual plot.

**(iv).** To predict future values using the models determine.

**(v).** To visualize data whose meaning in one sense is very different from another sense.

**(vi).** To highlight a few applications of bivariate data.

## Introduction

A times a research study may involve two variables and the researcher many be interested in determining the relationship between the two variables. Such data is referred to as bivariate data. In some cases, one variable may sort to explain the results of the other variable. To graphically study and analyze such kind of data, a scatter plot is very important to consider. From the plot, we can even predict future responses based on the behaviors of the data. In this lesson, we are going to look into the details of such data, data having two variables.

# Specific Vocabulary for Tutoring Session

### A Response and Explanatory Variables

It is a main variable under study in any give research study while an explanatory variable is a variable that tries to explain the results of the response variable.

### A Scatter Plot

It is a graph that shows a relationship between two variables based on the data given.

### Lurking Variable

It is a variable that is not the response variable not the explanatory variable in any given statistical research however, its inclusion would  significant when explaining the relationship between the variables  used in the study.

### An outlier

It is a point that does not follow the trend of other points in a plot

# Direct Instruction: Modeling for You

### Explanatory and Response Variables

A response variable is a main variable under study in any give research study. It can be identifies from the question of study. For instance, if the study is about the number of number of four-bedroom houses in every square kilometer of a town, then the response variable will be the number of such houses. In some cases, the question of study may be general to an extent that there may be a no response variable. For instance, the question study; study the causes and effects of land degradation in an area, does not have a response variable. This is because, there is no specific response that can be controlled in the study.

The explanatory variable is a variable that tries to explain the results of the response variable. For the case of the study question above, the response variable will be to determine the number of low income earners in the population. This would explain the response because, when the population has very many low income earners in an area, there is high chances that they will be very few four-bedroom houses since most of them cannot afford them.

## Question 1

How do you think a response variable and its corresponding variable should be plotted on a graph? (which axis should have one variable and which one should have the other)

### Responses

A response variable represents the results that are determined by the explanatory variable, therefore, the former (response variable) may be viewed as dependent variable while the latter (explanatory variable) may be viewed as independent variable, however, this does not imply that explanatory variable is purely a dependent variable, it some cases, it may not be exclusively independent. Therefore, the response variable will be on the vertical axis awhile the explanatory variable would be on the horizontal axis.

## Graphical representation of bivariate data: Scatter Plot

### Constructing a scatter plots

A **scatter plot** is a graph that shows a relationship between two variables based on the data given.

To construct a scatter plot, among the two variables provided, the dependent and the independent variable ( or more precisely, the response and the explanatory variables) are identified then the points plotted using the ordered pairs (explanatory variable, response variable).

### Illustration

Draw a scatter plot to represent the following data

| $X$ | 1 | 2 | 3 | 4 | 5 | 6 | 7 |
|---|---|---|---|---|---|---|---|
| $Y$ | 10 | 13 | 15 | 10 | 14 | 17 | 18 |

Since $x$ and $y$ are independent and dependent variables respectively, the ordered pairs will be in the form $(x,y)$.

Upon plotting, we have the following

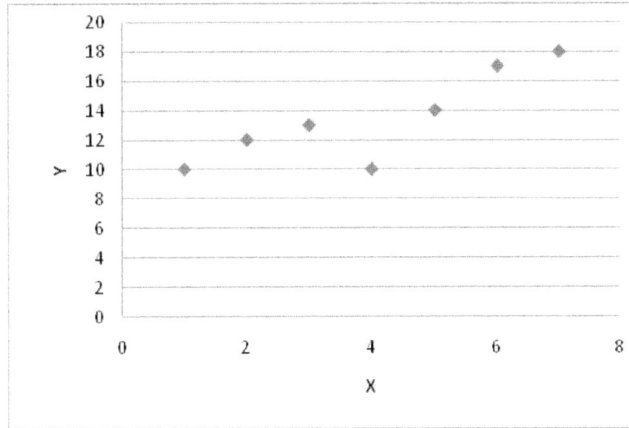

## Interpreting scatter plots

The scatter plot in most cases is aimed at showing a linear relationship between the variables plotted. The line is drawn to show this linear relationship is called the **line of best fit or the regression line**. The line would always appear to be at the center of all the points of the plot.

### Linear And Non Linear Relationships

When the plots represents a line as shown in the first diagram, we say that the **relationship is linear** though it has one outlier. An outlier is a point that does not follow the trend of other points in a plot. When the plots represents a curve like in the second diagram, we say that the relationship between the two variables is **non linear**.

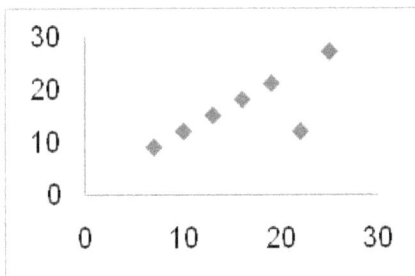

Linear relationship

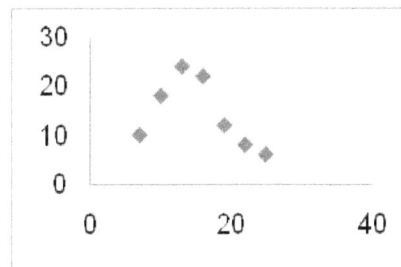

Non-linear relationship

### Interpretation Based on Linearity

When two variables have any sort of relationship, we say the variables are correlated. Thus, for the case above, the first graph are linearly correlated.

When a large percentage of the points are closer to the regression line, we say that the variables are strongly correlated and when a few are closer to it, the variables are weakly correlated. When the points are distributed all over the plot area, we say that the variables are not correlated ( no correlation).

The slope of the regression line determines the type of correlation, that is, positive or negative.

When the slope of the line is negative, then the variables are negatively correlated. When slope is positive, the variables are positively correlated.

The following diagrams shows different characterizations of a scatter plot.

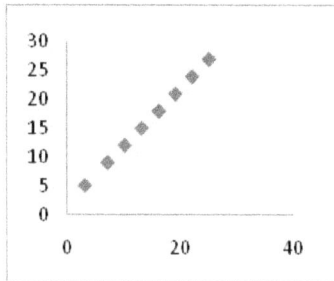

Maximum positive correlation

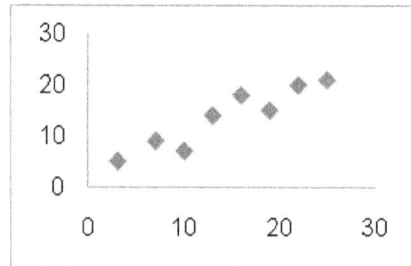

strong positive correlation

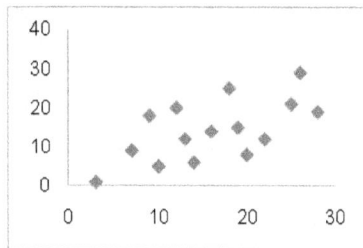

weak positive correlation

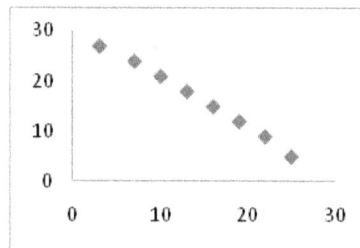

Maximum negative correlation

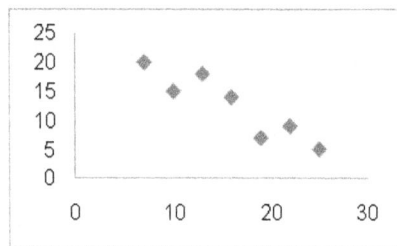

strong negative correlation

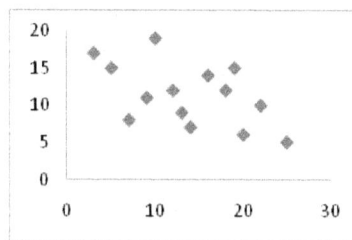

weak negative correlation

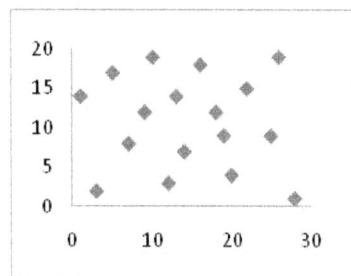

No correlation

## Question 2

Generate a scatter plot and describe the correlation of the following data

| $X$ | 1 | 2 | 3 | 4 | 5 | 6 | 7 | 8 | 9 |
|---|---|---|---|---|---|---|---|---|---|
| $Y$ | 5 | 1 | 11 | 9 | 13 | 14 | 15 | 13 | 16 |

### Responses

Putting $x$ values and $y$ values on the horizontal and vertical axes respectively, we have the following graph

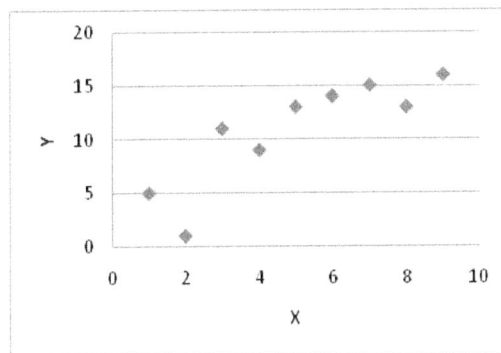

Most points are  not on but closer to the regression line, hence the correlation is strong. Since the regression line has a positive slope, the data has a strong positive correlation.

## Question 3

A survey was done to establish the relation between the rainfall amounts received in an area and the rate of deforestation.  The  rate of deforestation was determined by comparing the total forest cover of an area with respect to the forest cover in the previous year. The table below shows the results.

Identify the response and the explanatory variable. Plot a scatter plot and explain the results.

| Rainfall amount (in) | 1 | 2 | 3 | 5 | 5 | 6 | 7 | 7 | 8 |
|---|---|---|---|---|---|---|---|---|---|
| Deforestation rate (%) | 8 | 9 | 7 | 7 | 3 | 1.5 | 1 | 1 | 0.5 |

The rainfall amount received in an area is influenced by the vegetation among other things, therefore, the change in the vegetation cover affects the amount of rainfall received. Thus, the response variable is the rainfall amount while the explanatory variable is the rate of deforestation.

Thus, the scatter plot is plotted using the ordered pairs of the form (deforestation rate, rainfall amount)

The plot will be as one shown below.

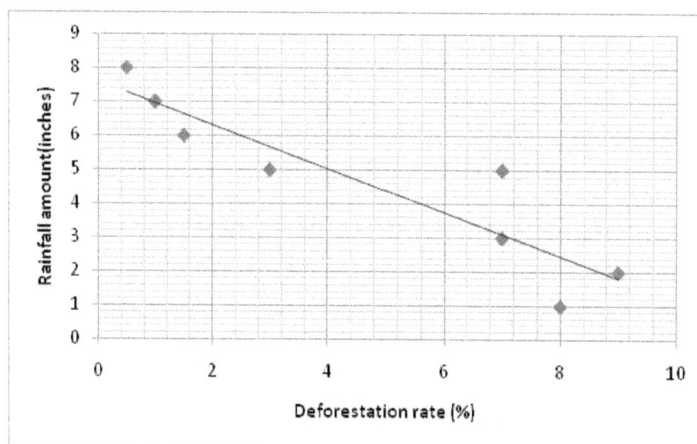

The line of best fit has a negative slope and the points are closer to it. Hence the rainfall amount and the deforestation rate are strongly and negatively correlated. The correlation is negative because the increase in deforestation reduces forest cover that interns decreases the rainfall amount received in an area due to reduced moisture released in the atmosphere.

## Correlation coefficient

The correlation coefficient, r, shows the strength and the nature of the correlation between the two variables. It takes values between and including  -1 and 1. It is also called the **Pearson product moment correlation coefficient**.

When the correlation is maximum positive, then r =1 and when it maximum negative, r  = -1. This implies that all the points lie on the regression line. These two cases are also referred to as **perfect correlation**.

When the correlation is strongly positive, the r is positive and closer to 1 and when it is strongly negative, r is closer to -1. This happens when most of the points are closer to the regression line.

The negative value of r implies that the variables are inversely proportional, that is, the increase of one variable leads to the decrease of another. When the value of r is positive, it implies that the variables are directly proportional, that is, the increase of variable leads to the increase of the other.

The correlation coefficient of two variables $X$ and $Y$ is given by

$$r = \frac{n\sum x_i y_i - \sum x_i \sum y_i}{\sqrt{n\sum x_i^2 - (\sum x_i)^2}\sqrt{n\sum y_i^2 - (\sum y_i)^2}}$$

# Guided Instruction: Working With You

## Question 4

A class teacher would like to determine the relationship between the height and the weight of his students. The table below shows the

| Height(in) | 57 | 58 | 58 | 59 | 59 | 60 | 62 | 63 |
|---|---|---|---|---|---|---|---|---|
| Weight (lbs) | 97 | 95 | 99 | 110 | 105 | 98 | 104 | 105 |

Determine and explain the correlation coefficient the coefficient

## Responses

We come up with a table and compute the values shown in the following table

Summations

| | | | | | | | | | |
|---|---|---|---|---|---|---|---|---|---|
| Height(in) , $x$ | 57 | 58 | 58 | 59 | 59 | 60 | 62 | 63 | 476 |
| Weight (lbs), $y$ | 97 | 95 | 99 | 100 | 103 | 98 | 104 | 105 | 801 |
| $xy$ | 5529 | 5510 | 5742 | 5900 | 6077 | 5880 | 6448 | 6615 | 47701 |
| $x^2$ | 3249 | 3364 | 3364 | 3481 | 3481 | 3600 | 3844 | 3969 | 28352 |
| $y^2$ | 9409 | 9025 | 9801 | 10000 | 10609 | 9604 | 10816 | 11025 | 80289 |

From the table, we have

$$\sum x_i = 476, \sum y_i = 801, \sum x_i y_i = 47701, \sum x_i^2 = 28352, \sum y_i^2 = 80289, n = 8$$

We now substitute the values in the following formula

$$r = \frac{n\sum x_i y_i - \sum x_i \sum y_i}{\sqrt{n\sum x_i^2 - (\sum x_i)^2}\sqrt{n\sum y_i^2 - (\sum y_i)^2}}$$

$$r = \frac{(8\times 48409) - (476\times 813)}{\sqrt{(8\times 28352) - (476)^2}\sqrt{(8\times 82805) - (813)^2}} = 0.8031$$

$r = 0.8031$ which is closer to 1

There is a strong positive correlation between the height and the weight of the students

## Coefficient of determination

This is a ratio that shows how well the fitted regression represents the data since it is a ratio of explained variations of one variable with respect to the total variation. From this, it is better to determine if the linear regression model determined is good for determining values within and outside the data given. The fact that it shows the variations in the response variable that is explained, it is important in cases where more than one explanatory variable is used. From the individual Coefficient of determinations of various explanatory variable, one is able to see which variable is more relevant when using the values to predict the data. When its value is high, then the variable is most significant in predicting values using the given model.

It is given by   where $r$ is the correlation coefficient.

In the question 4 above, the coefficient of determination will be $r^2 = 0.8031^2 = 0.6449$

Since the square of a number is always positive, the coefficient of determination takes values between and including 0 and 1   $0 \le r^2 \le 1$

## Question 5

A teacher wishes to establish the relationship between the performance in class and the amount of pocket money given to students. The therefore, get a sample by random sampling and interview them then get their average performance in the previous term to analyze the two. The table below shows the results.

| Amount of pocket money ($) | 100 | 120 | 130 | 150 | 200 | 240 | 240 | 260 |
|---|---|---|---|---|---|---|---|---|
| Performance (100%) | 80 | 73 | 64 | 76 | 60 | 72 | 64 | 81 |

Determine and explain the coefficient correlation and the coefficient of determination

We come up with a table and compute the values shown in the following table

**Summation**

| Amount of pocket money ($), $x$ | 100 | 120 | 130 | 150 | 200 | 240 | 240 | 260 | 1440 |
|---|---|---|---|---|---|---|---|---|---|
| Performance (100%), $y$ | 80 | 73 | 64 | 76 | 63 | 72 | 59 | 70 | 557 |
| $xy$ | 8000 | 8760 | 8320 | 11400 | 12600 | 17280 | 14160 | 18200 | 98720 |
| $x^2$ | 10000 | 14400 | 16900 | 22500 | 40000 | 57600 | 57600 | 67600 | 286600 |
| $y^2$ | 6400 | 5329 | 4096 | 5776 | 3969 | 5184 | 3481 | 4900 | 39135 |

From the table, we have
$$\sum x_i = 1440, \sum y_i = 557, \sum x_i y_i = 98720, \sum x_i^2 = 286600, \sum y_i^2 = 39135, n = 8$$

We now substitute the values in the following formula

$$r = \frac{n\sum x_i y_i - \sum x_i \sum y_i}{\sqrt{n\sum x_i^2 - \left(\sum x_i\right)^2}\sqrt{n\sum y_i^2 - \left(\sum y_i\right)^2}}$$

$$r = \frac{(8\times 98720)-(1440\times 557)}{\sqrt{(8\times 286600)-(1440)^2}\sqrt{(8\times 39135)-(557)^2}} = 0.8031$$

$r = 0.4939$ which is not closer to 1

There is a weak positive correlation between the performance of the students and the amount of pocket money given

$r^2 = 0.4939^2 = 0.2439 = 24.9\%$

This implies that the explanatory variable, the amount of pocket money given to the student can only contribute be attributed to 24.39% positively to the performance of the students.

## Calculating, Graphing and Interpreting Residual Plots

Residual is a the vertical difference between the plotted point and the regression line with respect to a particular value on the horizontal variable. They are abbreviated as e.

Residuals = measured value – predicted value

Where the predicted value is the value determined by the regression line and the measured value is the value in got from the study.

Below is an illustration.

Consider the case below the data and its corresponding scatter plot below

**Residual** is a the vertical difference between the plotted point and the regression line with respect to a particular value on the horizontal variable. They are abbreviated as e.

Residuals = measured value – predicted value

Where the predicted value is the value determined by the regression line and the measured value is the value in got from the study.

Below is an illustration.

Consider the case below the data and its corresponding scatter plot below

| $X$ | 10 | 50 | 30 | 70 | 90 |
|---|---|---|---|---|---|
| $Y$ | 10 | 20 | 30 | 38 | 50 |

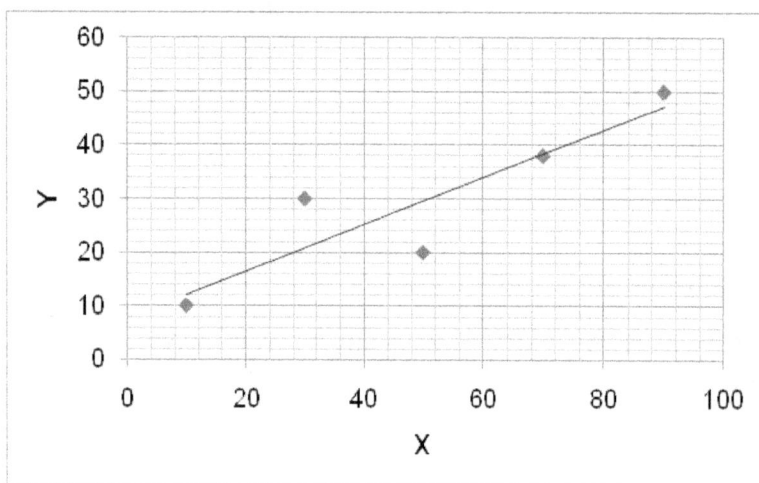

From the graph, the residuals will be, $10 - 12 = -2$; $30 - 20 = 10$; $20 - 30 = -10$, $38 - 38 = 0$; $50 - 47 = 3$

To get the residual plot, the residuals are plotted against the values of the independent variable (on horizontal axis).

The table will be

| X | 10 | 50 | 30 | 70 | 90 |
|---|----|----|----|----|----|
| e | −2 | 10 | −10 | 0 | 3 |

**Residual plot**

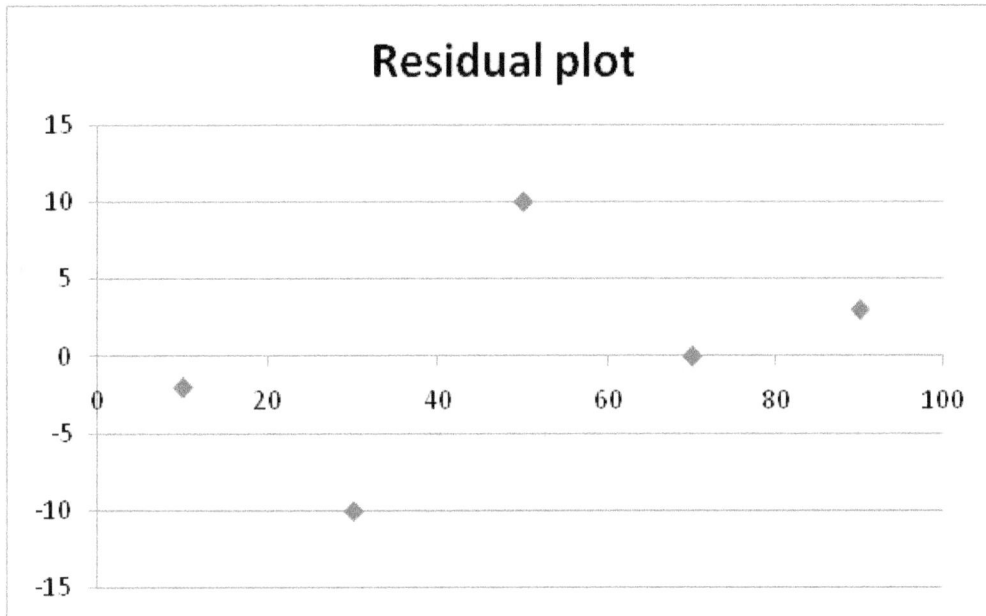

*Interpreting a residual plot*

The residual plot tells us the best method for representing the given data.

When the points in a residual plot are uniformly or randomly distributed around the line, then the linear model is the best for representing the data.

In the case above, the points are almost evenly distributed around the horizontal line, hence the linear model is the best for representing the data.

When the points of the residual plot forms a curve, then a non linear model is the best to represent the data.

Computational method of determining the residuals.

In most cases, the graph may to lead to accurate calculations of the residuals. Therefore, the sure method is by use of the regression model. The regression model that we will use in this case is the linear model.

Since it is linear, it is given by

$y = a + bx + e$ where $e$ is the residuals

The fitted regression line is $\hat{y} = \hat{a} + \hat{b}x$ where $\hat{b} = \dfrac{\sum(x-\bar{x})(y-\bar{y})}{\sum(x-\bar{x})^2}, \hat{a} = \bar{y} - \hat{b}\bar{x}$

We may also use an equivalent formula for $\hat{b}$

$$\hat{b} = \frac{\sum x - n\bar{x}\bar{y}}{\sum x^2 - n(\bar{x})^2}$$

The residuals are therefore given by the formula $e = y - \hat{y}$

## Question 6

**Generate a residual plot based on the following data.**

| X | 1 | 4 | 6 | 11 | 12 | 14 | 15 | 18 | 20 |
|---|---|---|---|----|----|----|----|----|----|
| Y | 17 | 18 | 15 | 17 | 23 | 30 | 19 | 22 | 27 |

## Responses

First, we determine the regression line model this is done by calculating the values of $\hat{b}$ and $\hat{a}$

| $n =$ | 9.00 | | | | | | | | | sum | Mean |
|---|---|---|---|---|---|---|---|---|---|---|---|
| $x_i$ | 1.00 | 4.00 | 6.00 | 11.00 | 12.00 | 14.00 | 15.00 | 18.00 | 20.00 | 101 | 11.222 |
| $y_i$ | 17.00 | 18.00 | 15.00 | 17.00 | 23.00 | 30.00 | 19.00 | 22.00 | 27.00 | 188 | 20.889 |
| mean of $x$ | 11.22 | 11.22 | 11.22 | 11.22 | 11.22 | 11.22 | 11.22 | 11.22 | 11.22 | 101 | |
| 9 | 20.89 | 20.89 | 20.89 | 20.89 | 20.89 | 20.89 | 20.89 | 20.89 | 20.89 | 188 | |
| $a = x_i -$ mean of $x$ | (10.22) | (7.22) | (5.22) | (0.22) | 0.78 | 2.78 | 3.78 | 6.78 | 8.78 | 0 | |
| $b = y_i - 9$ | (3.89) | (2.89) | (5.89) | (3.89) | 2.11 | 9.11 | (1.89) | 1.11 | 6.11 | 0 | |
| $ab$ | 39.7530864 | 20.864 | 30.753 | 0.8642 | 1.642 | 25.309 | −7.136 | 7.6309 | 53.642 | 173.22 | |
| $a^2$ | 104.49 | 52.16 | 27.27 | 0.05 | 0.60 | 7.72 | 14.27 | 45.94 | 77.05 | 329.55 | |

From the table we have $\sum(x-\bar{x})(y-\bar{y}) = 173.2$, $\sum(x-\bar{x})^2 = 329.56$

Upon substitution into the formula

We have

$$\hat{b} = \frac{\sum(x-\bar{x})(y-\bar{y})}{\sum(x-\bar{x})^2}, \hat{a} = \bar{y} - \hat{b}\bar{x}$$

$$\hat{b} = \frac{173.22}{329.56} = 0.5256 \quad \hat{a} = \bar{y} - \hat{b}\bar{x} = 20.89 - (0.5256 \times 11.22) = 15$$

Therefore, we have

$$\hat{y} = \hat{a} + \hat{b}x = 15 + 0.5256x \quad \hat{y} = 15 + 0.5256x$$

The residuals are given by $e = y - \hat{y}$

Using the linear regression model $\hat{y} = 15 + 0.5256x$ we calculate the values of $\hat{y}$ for each value of $x$.

Thus, we come up with the table

| $x_i$ | 1.00 | 4.00 | 6.00 | 11.00 | 12.00 | 14.00 | 15.00 | 18.00 | 20.00 |
|---|---|---|---|---|---|---|---|---|---|
| $\hat{y}$ | 15.53 | 17.10 | 18.15 | 20.78 | 21.31 | 22.36 | 22.88 | 24.46 | 25.51 |
| $y_i$ | 17.00 | 18.00 | 15.00 | 17.00 | 23.00 | 30.00 | 19.00 | 22.00 | 27.00 |
| $e^i = y_i - \hat{y}$ | (1.47) | (0.90) | 3.15 | 3.78 | (1.69) | (7.64) | 3.88 | 2.46 | (1.49) |

The residual plot is got by plotting the values of $e\_i$ against the values of $x\_i$, thus, we have

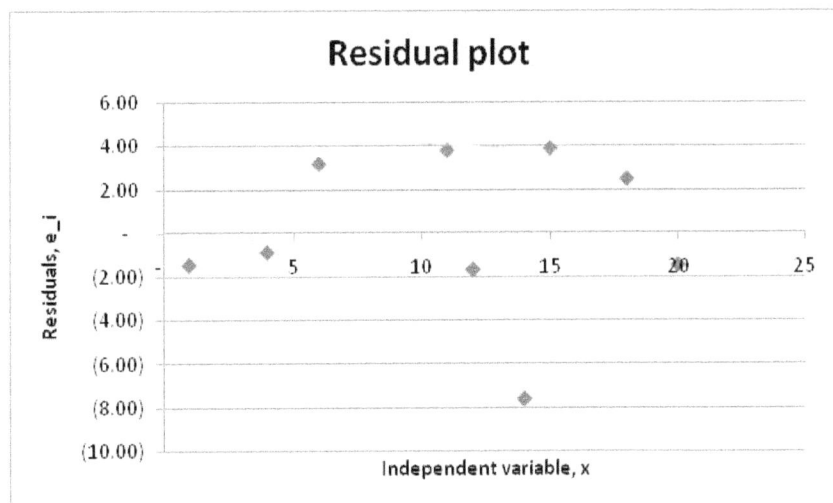

## Influencing points

An influencing point is a point in the data whose presence or absence significantly affect the regression line and model consequently, the coefficient of determination and the correlation coefficient. It does depends on the values of both independent and dependent variable. It is a bit different from the outlier point. The outlier point is a point that does not conform to the trend of other points in the data. It is depends on the dependent variable only.

Therefore, a point many be an outlier but fails to be an influential point in the data.

## Question 7

A survey was done to investigate between the length of footstep and height of an adult person. The table below shows the result.

| Height, inches | 56 | 58 | 60 | 60 | 65 | 65 | 66 | 67 | 68 |
|---|---|---|---|---|---|---|---|---|---|
| Footstep's Length, inches | 25 | 26 | 26 | 25 | 27 | 27 | 28 | 28 | 33 |

Identify the outlier and determine if it is an influencing point.

## Responses

We first determine plot the graph and determine linear regression models using both points and one without the outlier.

Upon plotting, we get the following data:

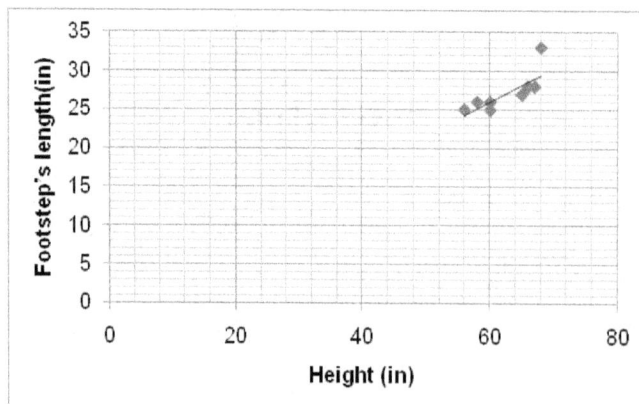

From the graph, the last point ,(68,33) appears not to follow the trend of the other hence it is an outlier point.

The determine the regression model using the formula

$$\hat{b} = \frac{\sum xy - n\bar{x}\bar{y}}{\sum x^2 - n(\bar{x})^2} \qquad \hat{a} = \bar{y} - \hat{b}\bar{x}$$

$n =$ 9                                                                 sum     Mean

| Height (in), $x_i$ | 56 | 58 | 60 | 60 | 65 | 65 | 66 | 67 | 68 | 565 | 62.77778 |
|---|---|---|---|---|---|---|---|---|---|---|---|
| Footstep's Length(in) $y_i$ | 25 | 26 | 26 | 25 | 27 | 27 | 28 | 28 | 33 | 245 | 27.22222 |
| $(x_i)^2$ | | 3136 | 3364 | 3600 | 3600 | 4225 | 4225 | 4356 | 4489 | 4624 | 35619 | |
| $(x_i)(y_i)$ | | 1400 | 1508 | 1560 | 1500 | 1755 | 1755 | 1848 | 1876 | 2244 | 15446 | |

Thus, we have $\sum xy = 15446$, $\sum x^2 = 35619$, $\bar{x} = 62.78$, $\bar{y} = 27.22$

Upon substitution in $\hat{b} = \dfrac{\sum xy - n\bar{x}\bar{y}}{\sum x^2 - n(\bar{x})^2}$ we have

$$\hat{b} = \frac{\sum xy - n\bar{x}\bar{y}}{\sum x^2 - n(\bar{x})^2} = \frac{15446 - (9 \times 62.78 \times 27.22)}{35619 - (9 \times 62.78^2)} = 0.437$$

$$\hat{a} = \bar{y} - \hat{b}\bar{x} = 27.22 - (0.4371 \times 62.78) = -0.2148$$

The model is $y = \hat{a} + \hat{b}x = -0.2148 + 0.437x$

We also determine the model without including the outlier:

$n =$ 8                                                                 sum     Mean

| Height (in), $x_i$ | 56 | 58 | 60 | 60 | 65 | 65 | 66 | 67 | 497 | 62.125 |
|---|---|---|---|---|---|---|---|---|---|---|
| Footstep's Length (in) $y_i$ | 25 | 26 | 26 | 25 | 27 | 27 | 28 | 28 | 212 | 26.5 |
| $(x_i)^2$ | 3136 | 3364 | 3600 | 3600 | 4225 | 4225 | 4356 | 4489 | 30995 | |
| $(x_i)(y_i)$ | 1400 | 1508 | 1560 | 1500 | 1755 | 1755 | 1848 | 1876 | 13202 | |

$\sum xy = 13202$, $\sum x^2 = 30995$, $\bar{x} = 62.13$, $\bar{y} = 26.5$

$$\hat{b} = \frac{\sum xy - n\overline{x}\,\overline{y}}{\sum x^2 - n(\overline{x})^2} = \frac{13202 - (8 \times 62.13 \times 26.5)}{30995 - (8 \times 62.13^2)} = 0.256$$

$$\hat{a} = \overline{y} - \hat{b}\overline{x} = 26.5 - (0.256 \times 62.13) = 10.59$$

Thus $y = \hat{a} + \hat{b}x = 10.59 + 0.256x$

There is a significant difference between the constants of the first model, -0.2148 and the second model, 10.59. The slope of the models also changes by -0.181. The changes are significant and intern affects the correlation coefficient and the coefficient of determination significantly. Thus, the point (68,33) is is an influencing point

## Lurking variable

In most of the examples and the questions that we have used in this lesson, we were considering only two variables. Sometimes, the conclusion made is not reliable since there may be some background factors that would explain the situation and thus the relation between the two variables involved in analysis.

For instance, consider the case question whose data and scatter plot are

| Rainfall amount(in) | 1 | 2 | 3 | 5 | 5 | 6 | 7 | 7 | 8 |
|---|---|---|---|---|---|---|---|---|---|
| Deforestation rate (%) | 8 | 9 | 7 | 7 | 3 | 1.5 | 1 | 1 | 0.5 |

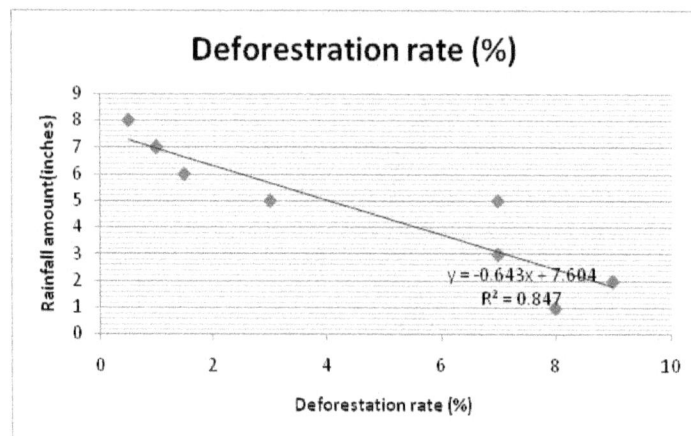

$y = -0.643x + 7.604$
$R^2 = 0.847$

The amount of rainfall is inversely proportional to the deforestation rate and the correlation is strong. The data implies that when we reduce deforestation, the rain would increase which in actual sense is not one of the major factors. There are some major factors in the background that would explain the response variable. One of them may be the number of major forest(s) in the area. This affects both the rainfall and the deforestation rate. When there are major forests in the area, the amount of rainfall will be high.

Likewise, the deforestation rate would not increase significant since the rate would be low as a result of many trees in the area. Therefore, this variable, the number of large forest(s) in the area is the lurking variable. Therefore, the lurking variable is a variable that is not the response variable not the explanatory variable in any given statistical research however, its inclusion would significant when explaining the relationship between the variables used in the study.

## Extrapolation

This is the prediction of values that are not within the data given. It is possible that given a data from a field study, one would predict the future values by observing the trend or using a model. In our case, we will use the linear regression model to predict feature values.

The fitted model is, given that value of $x$, we can determine the corresponding $y$ estimate value

## Question 8

A fruit vendor relocated to a new location, a place where he is not sure about the market trends. She therefore recorded the total amount of sales made within his first 9 days at the new location.

| Days | 1 | 2 | 3 | 4 | 5 | 6 | 7 | 8 | 9 | 10 |
|------|-----|-----|-----|-----|-----|-----|-----|-----|-----|-----|
| Sales($) | 105 | 130 | 155 | 165 | 205 | 215 | 225 | 220 | 230 | 235 |

Estimate the total sales in the 12th day.

We determine the linear model using the formula $\hat{y} = \hat{a} + \hat{b}x$ where $\hat{b} = \dfrac{\sum xy - n\bar{x}\bar{y}}{\sum x^2 - n(\bar{x})^2}$ and

$\hat{a} = \bar{y} - \hat{b}\bar{x}$

We therefore come up with a table shown below:

| $n =$ | | 10 | | | | | | | | | sum | mean |
|---|---|---|---|---|---|---|---|---|---|---|---|---|
| Days , $x$ | 1 | 2 | 3 | 4 | 5 | 6 | 7 | 8 | 9 | 10 | 55 | 5.5 |
| Sales(\$), $y$ | 105 | 130 | 155 | 165 | 205 | 215 | 225 | 220 | 230 | 235 | 1885 | 188.5 |
| $x^2$ | 1 | 4 | 9 | 16 | 25 | 36 | 49 | 64 | 81 | 100 | 385 | |
| $xy$ | 105 | 260 | 465 | 660 | 1025 | 1290 | 1575 | 1760 | 2070 | 2350 | 11560 | |

Thus $\sum x^2 = 385$, $\bar{x} = 5.5$, $\bar{y} = 188.5$, $\sum xy = 11560$

$\hat{b} = \dfrac{\sum xy - n\bar{x}\bar{y}}{\sum x^2 - n(\bar{x})^2} = \dfrac{11560 - (10 \times 5.5 \times 188.5)}{385 - (10 \times 5.5)} = 14.45$

$\hat{a} = \bar{y} - \hat{b}\bar{x} = 188.5 - (14.45 \times 5.5) = 109$

The model will be $\hat{y} = 109 + 14.45x$

To get the sales for the twelfth day, we take $x = 12$

Using the model, we have $\hat{y} = 109 + (14.45 \times 12) = 282.4$

The sales would be 282.4

## Linearizing data

In some cases, a linear regression may not be the best to describe the data. This happens when the points of a residual plot shows a trend that that is not linear, a curve. When the trend shows a quadratic curve, logarithmic among others, then the data can be modified using the following transformations of the variables so as to linearize it. To use the transformation, one has to know the king of curve the residual plot depicts.

| Original curve | Equation | What to plot |
|---|---|---|
| Quadratic | $y = ax^2 + bx + c$ | $(x, \sqrt{y})$ |
| Hyperbolic | $y = \dfrac{a}{b+x}$ | $(x, \dfrac{1}{y})$ |
| Exponential | $y = ab^x$ | $(x, \ln y)$ |
| Power | $y = ax^b$ | $(\ln x, \ln y)$ |

Once linearized, data can be analyzed in the usual way as a linear model.

## Simpson's Paradox

Simpson paradox is a statistical description of data where there is a reverse of inferences made about a particular thing or idea based on different conditions. A times one may make a few observations and come up with a general description of the variables yet the description changes when significantly a number of observations and done and systematically analyzed.

For instance, when may be interested about the gender representation in the number of students taking nutrition course. On taking a small sample such as 10 people, one may get that the number of ladies is 4 while that of that of male may be 6. This implies that the majority of the students taking nutrition course are male. However, there may be a reverse of this conclusion when all the students taking nursing are involved. For instance, when their total number is 68 where the female students are 42 and that of males is 26. By considering all students, we find that the female are the majority. This situation of reversal in conclusion about the data after considering a large number of sample is called the Simpson's paradox.

# Video Suggestions for Tutoring Sessions

Please conduct a search on either YouTube or Teacher Tube to find appropriate videos for this lesson. Below are some suggested title searches:

1. Use of Scatter Plot Method in Agriculture

2. Use of Simpson's Paradox in Analyzing Scores in a Game

3. Use of Scatter Plot in Making Decisions in the Business field

# Independent Instruction: Working on Your Own

## Questions

1. Generate a scatter plot for the following data

| h | 2 | 4 | 5 | 7 | 8 | 10 | 12 |
|---|---|---|---|---|---|----|----|
| K | 0.3 | 0.41 | 0.42 | 0.61 | 0.57 | 0.7 | 0.71 |

2. What could be the coefficient of determination of the following data

| x | 1 | 3 | 4 | 6 | 8 | 9 | 11 | 14 |
|---|---|---|---|---|---|---|----|----|
| y | 5 | 5 | 7 | 9 | 6 | 8 | 10 | 9 |

3. Describe the correlation of the following variables

| s | 42 | 52 | 63 | 67 | 73 | 79 | 82 |
|---|----|----|----|----|----|----|----|
| r | 123 | 117 | 118 | 116 | 120 | 117 | 116 |

**1.** We plot the values of $k$ against $h$

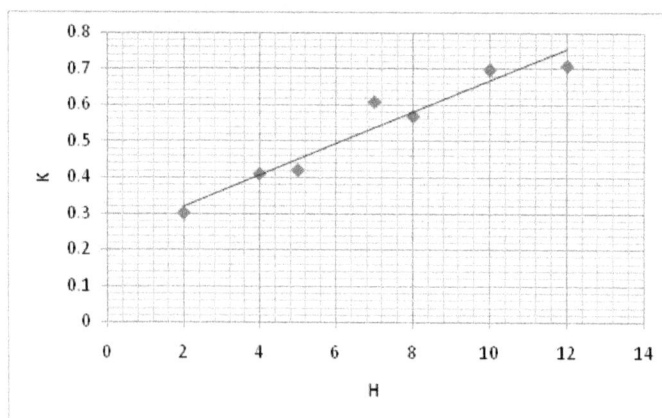

**2.** We come up with the table that will enable us determine the value of $r$ where

$$r = \frac{n\sum x_i y_i - \sum x_i \sum y_i}{\sqrt{n\sum x_i^2 - (\sum x_i)^2}\sqrt{n\sum y_i^2 - (\sum y_i)^2}}$$

| $x$ | 1 | 3 | 4 | 6 | 8 | 9 | 11 | 14 | sum 56 |
|---|---|---|---|---|---|---|---|---|---|
| $y$ | 5 | 5 | 7 | 9 | 6 | 8 | 10 | 9 | 59 |
| $xy$ | 5 | 15 | 28 | 54 | 48 | 72 | 110 | 126 | 458 |
| $x^2$ | 1 | 9 | 16 | 36 | 64 | 81 | 121 | 196 | 524 |
| $y^2$ | 25 | 25 | 49 | 81 | 36 | 64 | 100 | 81 | 461 |

$$r = \frac{n\sum x_i y_i - \sum x_i \sum y_i}{\sqrt{n\sum x_i^2 - (\sum x_i)^2}\sqrt{n\sum y_i^2 - (\sum y_i)^2}} = \frac{(8\times 458)-(6\times 9)}{\sqrt{(8\times 524)-6^2}\sqrt{(8\times 461)-9^2}} = 0.7691$$

Coefficient of determination is

$$r^2 = 0.7694^2 = 0.592$$

**3.** To describe the correlation, we make the scatter plot of the data

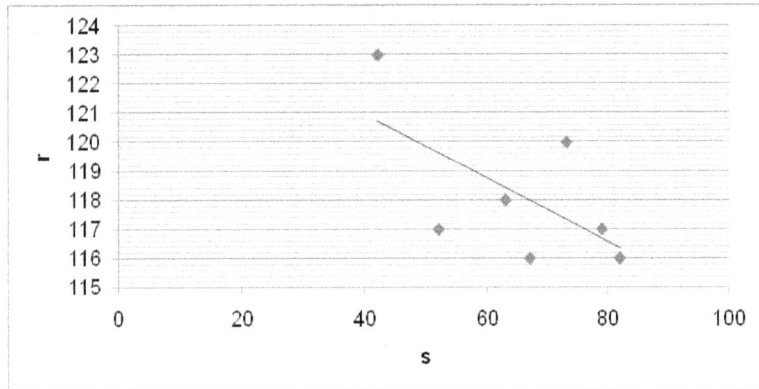

The slope of the lie is negative and the data points are sparsely distributed.

The absolute value of the slope is very little, less than 0.5. Therefore, the two variables, $r$ and $s$ are negatively weakly correlated.

# Mini-Assessment

1. A study was done to determine the relationship between the rainfall amounts in the area and the rate if soil erosion in an area. Identify the possible lurking variable of this study.

   **A.** Rate of soil erosion

   **B.** The climate of the area

   **C.** The human activities such as deforestation

   **D.** The amount of rainfall

   **E.** The population of the area

2. Identify the plot that best describes a linear relation

**A.**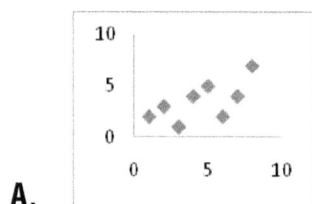

**B.**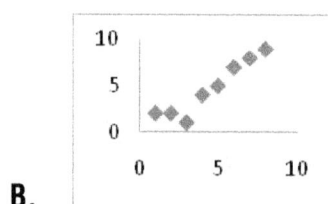

**C.**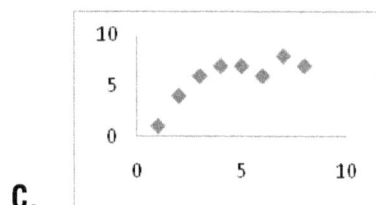

**D.**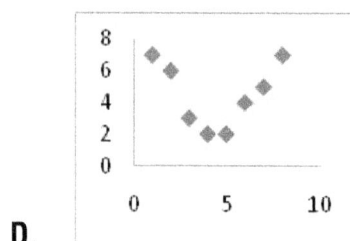

**E.**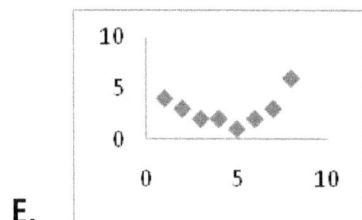

3. A researcher wants to know how much his regression model fits the data. What is the best statistics should he calculate.

   **A.** Standard Error       **B.** Correlation coefficient       **C.** Mean
   **D.** Coefficient of determination       **E.** Slope of the model

4. A wildlife research based organization would like to establish the relationship between the number of zebras and lions in a park. He moves from one park to another collecting the information. Given that he collects the information below, describe how the population of the two animals are correlated?

| No. of Zebra | 127 | 136 | 155 | 156 | 180 | 222 | 234 | 240 | 251 | 265 |
|---|---|---|---|---|---|---|---|---|---|---|
| No. of Lion | 87 | 88 | 100 | 98 | 103 | 105 | 120 | 123 | 121 | 132 |

5. A research study shows that the correlation between the student performance and the quality of teacher's delivery is a perfect positive correlation. Explain what this mean and criticize or explain why it is true.

6. The table below shows Regina's monthly expenses. Determine the model that describes her expenses.

| Month | 1 | 2 | 3 | 4 | 5 | 6 | 7 | 8 |
|---|---|---|---|---|---|---|---|---|
| Expenses ($) | 452 | 469 | 510 | 492 | 489 | 503 | 506 | 521 |

**7.** Davis a High school teacher and Joel an elementary teacher have been keeping the records of expenditure. They decide to compare the data and determine if there is any sort of relationship. Determine the correlation coefficient of their data.

| Davis expenditure ($) | 928 | 940 | 932 | 938 | 942 | 969 | 960 | 953 | 941 |
|---|---|---|---|---|---|---|---|---|---|
| Joel's expenditure ($) | 782 | 701 | 790 | 853 | 822 | 828 | 813 | 798 | 797 |

# Mini-Assessment Answers and Explanations

**1.** C

The lurking variable is a variable that is not part a response or an explanatory variable in the study, however, it helps in explaining the response variable and affects both two variables. The correct answer is the human activities such as deforestation

**2.** B

A linear relation must have the points making a trend similar to a straight line. The correct graph is the second one.

**3.** D

The statistic that shows how much his regression model fits the data is the coefficient of determination

**4.** We generate a scatter plot using the data

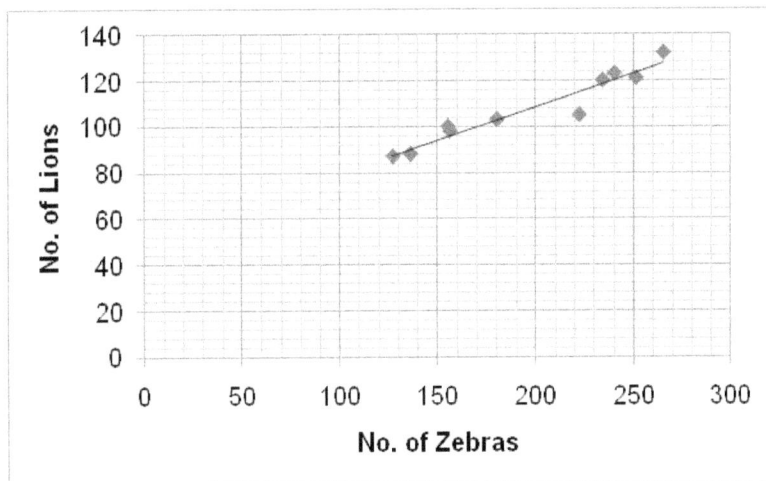

The graph has a positive slope with large value of the intercept. Therefore, the population of the two animals is strong positively correlated.

**5.** A perfect positive correlation shows that the slope of the regression line representing their relationship is 1. This means that the performance is wholly dependent on the quality of the teacher's delivery. This is not correct because there are other factors that affect students performance. These are preparation for the exams, the teaching and study facililities among others. When all these have to be incorporated, the contribution of the teachers' delivery would be less than 100% of all the contributions.

**6.** We come up with a table that would help us get the model

| $n =$ 8 | | | | | | | | | sum | mean |
|---|---|---|---|---|---|---|---|---|---|---|
| Month, $x$ | 1 | 2 | 3 | 4 | 5 | 6 | 7 | 8 | 36 | 4.5 |
| Expenses ($), $y$ | 452 | 469 | 510 | 492 | 489 | 503 | 506 | 521 | 3942 | 492.75 |
| $x^2$ | 1 | 4 | 9 | 16 | 25 | 36 | 49 | 64 | 204 | |
| $xy$ | 452 | 938 | 1530 | 1968 | 2445 | 3018 | 3542 | 4168 | 18061 | |

$$\hat{b} = \frac{\sum xy - n\overline{x}\,\overline{y}}{\sum x^2 - n(\overline{x})^2} = \frac{18061 - (8 \times 4.5 \times 492.8)}{204 - (8 \times 4.5^2)} = 7.666 \quad \hat{a} = \overline{y} - \hat{b}\overline{x} = 492.8 - (8 \times 4.5) = 456.8$$

The model is $y = 7.66x + 456.8$

**7.** 

| $n =$ 9 | | | | | | | | | sum |
|---|---|---|---|---|---|---|---|---|---|
| Davis expenditure ($), $x$ | 928 | 940 | 932 | 938 | 942 | 969 | 960 | 953 | 941 | 8503 |
| Joel's expenditure ($), $y$ | 782 | 701 | 790 | 853 | 822 | 828 | 813 | 798 | 797 | 7184 |
| $x^2$ | 861184 | 883600 | 868624 | 879844 | 887364 | 938961 | 921600 | 908209 | 885481 | 8034867 |
| $y^2$ | 611524 | 491401 | 624100 | 727609 | 675684 | 685584 | 660969 | 636804 | 635209 | 5748884 |
| $xy$ | 725696 | 658940 | 736280 | 800114 | 774324 | 802332 | 780480 | 760494 | 749977 | 6788637 |

$$r = \frac{n\sum x_i y_i - \sum x_i \sum y_i}{\sqrt{n\sum x_i^2 - \left(\sum x_i\right)^2}\sqrt{n\sum y_i^2 - \left(\sum y_i\right)^2}} = \frac{(9 \times 6788637) - (8503 \times 7184)}{\sqrt{(9 \times 8034867) - 8503^2}\sqrt{(9 \times 5748884) - 7184^2}} = 0.2983$$

The correlation is 0.2983

# Lesson Reflection

In this lesson, we have looked at the types of variables in bivariate data and saw how they can be represented graphically using scatter plot. To interpret the data, we have done regression analysis and determined some statistics that best describe the data. We have also looked at a way to determining if a linear regression is the best to analyze the data, this was done using a residual plot. We found that, when the data does not represent a linear relationship, we can transform it to linear by using a number of transformations. We have also looked at how to predict future values and finally looked the Simpson paradox and how the lesson can be applicable in real life situation.

# Lesson 4
# Randomness & Probability

## Lesson Description:

This lesson is designed to help students understand the concept of randomness and sample space. Students will have an opportunity to determine the probability of various events with different characteristics, along with applying the concept of probability in daily life. Please be sure to utilize the questions to help spark student engagement and cover the vocabulary that is associated with this specific tutoring session. For your own knowledge, sample responses have been provided to guide you as well.

## Learning Objective(s):

**(i).** To understand the concept of randomness and sample space.

**(ii).** To determine the probability of various events with different characteristics.

**(iii).** To see how we can apply the concept of probability to daily life.

## Introduction

We leave in a world in which no one can know what may happen next without relying on previous events. That is how nature is. However, most of the natural activities and events are important and some harmful thus necessitating their predictability. The fact that their occurrence cannot be determined by accuracy, we ray that happen at random. The study of these random events based of fractional representations is called probability. In this lessons, we will familiarize ourselves by the concepts of describing the certainty of occurrence of events with different characteristics.

# Specific Vocabulary for Tutoring Session

## Randomness

It refers to a situations where one is not certain when and if an activity or an event will happen.

## Probability

Is the ratio of the number of a particular events in an experiment to that of all possible outcomes.

## A Probability Model

It is a distribution that shows the probabilities of various outcomes in a given experiment.

## A set

A collection of items or activities with similar characteristics.

# Direct Instruction: Modeling for You

## Randomness and Probability of an Event

Randomness refers to a situations where one is not certain when and if an activity or an event will happen. Such activities described are referred to as random events. The occurrence of random events, however, can be predicted if observed over a long period of time since they have a regular distribution over time. Randomness is measured using probability.

$$\text{Probability of an event} = \frac{\text{Number of outcomes of the event}}{\text{Number of all oucomes}}$$

While investigating a statistical experiment where we are concerned about a certain event, then we talk about the number of all possible outcomes and the number of that particular events we are interested in. The number of all possible outcomes in an experiment is called the sample space. The ratio of the number of a particular events in an experiment to that of all possible outcomes is called probability.

It is sometimes denoted P(A) to mean the probability of an event A happening

Probability may also be expressed as a percentage. In the probability is $\frac{2}{5}$, then we can also express it as a percentage by multiplying it 100% to get $\frac{2}{5} \times 100\% = 40\%$

## Illustration

If a boy has six balls in which two are red and four are green and is interested in determining the probability of picking a red ball, this is how he goes about it. Since there are only six balls in total to pick from, the total number of outcomes is 6. The event is picking a red ball, hence the number of outcomes of the event is 2 since there are 2 red balls

The probability of selecting a red all will be $\dfrac{\text{Number of red balls}}{\text{Number of all balls}} = \dfrac{2}{6} = \dfrac{1}{3}$

## Question 1

Notice that in our illustration above, we have not provided the units for the probability of selecting a red ball, what could be the reason.

### Response

Probability is ratio of two similar events or activities where each one of them has similar units. Therefore, when dividing the number of outcomes of an event with the total number of outcomes, the units cancel out and we remains with a number with no units.

Therefore, probability does not have units.

## Question 2

A class teacher is in charge of a 20 students class. He wishes to pick a female student at random to be the class representative. If there are 8 boys in class, what is the probability of picking the female student?

### Response

The total number of students = 20

The number of boys = 8

The number of girls = 20 − 8 = 12

The event if picking a female student hence the number of students within he can pick is student is 12.

The probability of picking the female students will be

$\dfrac{\text{Number of female students}}{\text{Total number of students}} = \dfrac{12}{20} = \dfrac{3}{5} = 0.6$

## Dependence, Independence and Mutually Exclusive Events

Two events are referred to as dependent if the outcome of one event affects that outcome of the other. It that case, the probability of the event depend on that of the other event. For instance, the days being cloudy and a day being rainy are two deferent events but dependent. The event of raining depends on the event of being cloudy.

The probability that a days will be rainy is not the same as the probability that the day will be rainy given that it is cloudy.

In notation, the probability of dependent event is given as P(A/B) which is read as the probability of event A happening given that B happens. In a selection process without replacement, the subsequent processes are dependent upon the previous process.

Two events are referred to as independent if the outcome of one event does not affect the outcome of the other. Example, the event of a person visiting a part the event of a person being a fun of a soccer club are two independent events.

Two events are mutually exclusive if the occurrence of both two events cannot occur together. For instance, consider the event that a head and a tail occurs in a toss of a fair coin. The two events cannot occur at the same time, therefore, they are mutually exclusive events.

Mutually exclusive events should not be confused with independent event. In independent events, the occurrence of one does not affect the occurrence of the other therefore, there is a possibility that the two may occur at the same time. In mutually exclusive events, the two cannot occur at the same time.

In the example above about the event of a person being a fun of a particular soccer club and the event that a person visits a park can occur at the same time. This is because, we can have a soccer club fun visiting a park but the two are independent events. But since they can occur at the same time, they are not mutually exclusive events. Mutually exclusive events cannot be independent.

### Question 3

Alice has 40 blue and 56 white beads. He selects one blue bead then selects another one while making a necklace. What is the probability of selecting the second blue bead. Show that the probability of selecting the two bcads are dependent.

The total number of beads = 40 + 56 = 96

The number of blue beads = 40

The probability of selecting a blue bead for the first time = $\dfrac{40}{96} = \dfrac{5}{12}$

After the first blue bead is selected, the number of blue beads becomes = 40 − 1 = 39

The new total becomes 96 − 1 = 95

The values have been changes by the first selection hence the first event depends on the first one.

The probability of selecting a blue bead a second time will be $\dfrac{39}{95}$

## Probability Models

A probability model is a distribution that shows the probabilities of various outcomes in a given experiment. In this subtopic, we will discuss some of the basic properties of probability models. Probability is a number that varies between 0 and 1. When on is certain that something will occur, the probability of occurrence is 1. When one is certain that an event will not occur, then the probability of occurrence is zero. When the event is likely to happen, then probability is more than 0.5 or 50% and when an event is not likely to happen, the probability is less than 0.5 or 50%. When there are equal chances that an event will occur, the probability is 1.

Therefore, we come with a general inequality describing the probability of an event $X$ as $0 \le P(X) \le 1$

## Inclusion and Complementation

Consider a statistical experiment where the desired event is $X$. If the number of $X$ outcomes is a and the total number of outcomes is n, then the probability of event $A$ happening is $P(A) = \dfrac{a}{n}$

The event of $A$ not happening is called the complement of $A$ and is denoted $A^C$

The probability of A not happening is $P(A^c) = \dfrac{n-a}{n}$

Using some algebraic manipulations, we get that $P(A^c) = \dfrac{n-a}{n} = \dfrac{n}{n} - \dfrac{a}{n} = 1 - \dfrac{a}{n} = 1 - P(A)$

Therefore, the probability of an event not happening is 1 − probability of an event happening $P(A^c) = 1 - P(A)$

When two events can be described in terms of sets( a collection of items with similar characteristics) where the size of one set (say $A$) is greater than the other (say set $B$), we talk of set inclusion, that is; $B$ is a subset of $A$, $B \subseteq A$ in terms of symbols. Their probability will be related as $P(B) \leq P(A)$

The following is an illustration of probability on inclusion

Consider a situation where a science club has 8 members where 5 are female and 3 are male students. Then the set of, M, has fewer members than the set L, of female students. Thus we have the symbolic representation $M \subseteq L$

Probability of selecting a male is $P(M) = \dfrac{3}{8} = 0.375$

Probability of selecting a female student is $P(L) = \dfrac{5}{8} = 0.625$

Since $0.375 < 0.625$ then $P(M) \leq P(L)$

Therefore, If $M \subseteq L$ then $P(M) \leq P(L)$

## The Sum of the Probabilities

Given a statistical experiment the sum of all probabilities is 1. In the case above, the probability of an event A happening is $P(A)$ while that of A not happening is $P(A^C) = 1 - P(A)$. The sum of these two will be

$$P(A) + P(A^C) = P(A) + 1 - P(A) = P(A) - P(A) + 1 = 1$$

### Question 4

In school, 45% of the student likes darts game. What is the probability that a person chosen at random does not like darts.

### Response

The percentage of people whole likes darts , $D$, = 45% represents the fraction of people whole likes darts. This is similar to the probability of people whole likes darts. The probability of people who does not like darts will be the complement of $D$, that is

$$P(D) = 45\% = \frac{45}{100} \quad P(D^c) = 1 - P(D) = 1 - \frac{45}{100} = \frac{100}{100} - \frac{45}{100} = \frac{55}{100} = 0.55$$

The probability is 0.55

# Rules Of Probability

There are two major rules that are used in solving probability problems. These are addition and multiplication rules.

## Addition rules

The addition rule in probability sometimes implies the probability if the union of sets. The union of sets implies the another set having all members of all sets in the union. If A and B are two events, then the probability of $A$ or $B$ happening is $P(A \text{ or } B) = P(A) + P(B) - P(A \cap B)$.

The above formula is a general formula for all kind of events. However, when the events are mutually exclusive, that is the events cannot happen at the same time, then there is no common feature between the two sets, that is, $(A \cap B)$ has no element, $P(A \cap B) = 0$. Hence In such a case, we say that probability of event $A$ or $B$ happening where the two are mutually exclusive is $P(A \text{ or } B) = P(A) + P(B)$.

In most cases, this rule is simplified easier using diagrams of sets called the venn diagram.

When we have two sets, $A$ and $B$, then we have the following areas representing their intersection, union and complement.

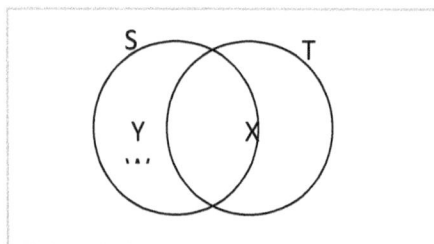

In the figure above, $S$ and $T$ are the main sets that are intersecting. $X$ is the intersection of $S$ and $T$ hence it contains elements that are common to both sets. $W$ is the set that has all elements that are not in $S$ and $Y$ is the set that has all elements that are not in $T$. Using venn diagram, we can easily visualize the situation and determine probabilities with ease.

# Direct Instruction: Working With You

## Question 5

A class is composed of 25 students where 16 likes drama and 6 likes both music and drama. Determine the probability that a student selected at random

**(i)** Likes music only

**(ii)** Likes music only or likes drams only

## Response

We can easily visualize this using venn diagram

We have two sets, those who like drama, $D = 16$

those who like music., $C = ?$

those who like both music and drama $A = 6$

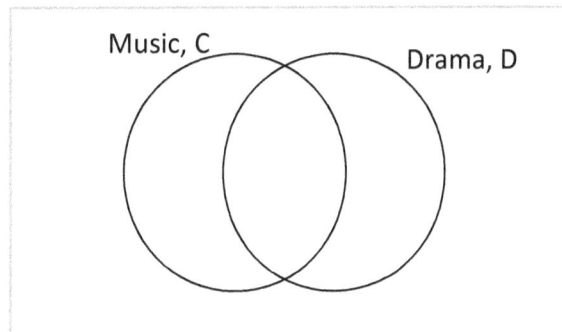

Music, C    Drama, D

Since $C = 16$, and $C = F + A$; $16 = F + 6$; $F = 10$

Those who likes music only are 10

Since to get the number of the whole class, we unite the two groups, We can get the number of those who likes drama only by subtracting the number of those who likes music from the number of all students.

That is the whole class, $25 = C + E$ or $25 = F + A + E$; since $C = F + A$. We have that $C = 16$,

Hence $25 = 16 + E$; $E = 9$

Probability of those who likes music only $= P(F) = \dfrac{10}{25} = 0.4$

Probability of those likes music only or likes drams only $= P(F \text{ or } E) = P(F) + P(E) + P(F \cap E)$

But $F$ and $E$ do not intersect hence their intersection is a null set (a set without an element). Thus $P(F \cap E) = 0$

Also $P(E) = \dfrac{9}{25}$

Thus, $P(For\ E) = \dfrac{10}{25} + \dfrac{9}{25} + 0 = \dfrac{19}{25} = 0.76$

We may also have a venn diagram having three or more sets. Consider the case below

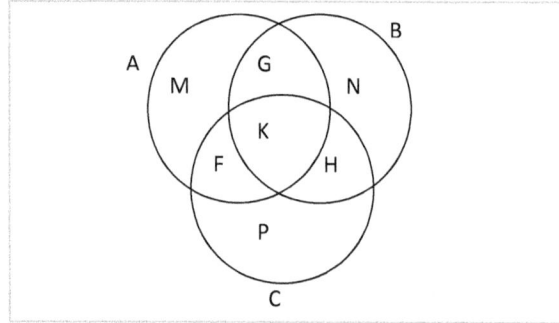

Here, the venn diagram has three main sets $A$, $B$ and $C$ that intersects. $K$ is a set that has all elements common to set $A$, $B$ and $C$. $F$, $H$ and $G$ are sets that have elements common to $A$ and $C$ only, $C$ and $B$ only and $A$ and $B$ only but are not common to all elements. In general, $F + K$, $H + K$ and $K + G$ are sets with elements that are common to $A$ and $C$ only, $C$ and $B$ only and $A$ and $B$ only. $M$, $N$ and $P$ contains elements that are in $A$ only, $B$ only and $C$ only respectively.

## Multiplication Rule

The multiplication of probabilities differ depending on the nature of the events; dependent or independent.

In this subtopic, we consider multiplication where the events are independent. The probability of an event $A$ and $B$ happening given that they are independent events is $P(A\ and\ B) = P(A)\ P(B)$.

### Question 6

40% of a population in town gets infected by malaria at least once a year while 45% gets infected by flue at least once a year. What is the probability that a person selected is not infected by both malaria and flue at least once per year?

We first list the individual probabilities

Let the probability that a person is infected by malaria at least once per year be $P(M)$

Let the probability that a person is infected by flu at least once per year be $P(F)$

Let the probability that a person is infected by malaria and flu at least once per year be $P(M \text{ and } F)$

$$P(M) = 40\% = \frac{40}{100} = 0.4, \quad P(F) = 45\% = \frac{45}{100} = 0.45 .$$

We now determine the probability $P(M \text{ and } F)$

The two events are independent hence

We determine the probability that a person is not infected by both infections

$P(\text{Not infected by malaria and flu}) = 1 - P(M \text{ and } L) = 1 - 0.18 = 0.82$

## Conditional Probability

This occurs when the occurrence of one event affects the occurrence of the other. In such situations, the events are said to be dependent. If the happening of an event $A$ dependent on $B$, then we write the probability of $A$ and $B$ happening together

$P(A \text{ and } B) = P(A) \, P(B/A)$ where $P(B/A)$ is read as probability of $B$ given $A$. $P(B/A)$ is the conditional probability.

Note that for the case where $A$ and $B$ are independent, the $P(B/A) = P(B)$ and $P(A/B) = P(A)$

## Question 7

A child has 9 yellow and 7 red marbles. He is instructed to pick two marbles, what is the probability that both are yellow or both are red.

**Response**

We list the notations

Let the probability that both are yellow be $P(YY)$ and the probability that both are red be $P(RR)$

YY implies that a yellow marbles is picked in the first trial and a yellow marble is picked in the second trial

Hence $P(YY) = P(Y \text{ and } Y)$ ; $P(RR) = P(R \text{ and } R)$

Since the probability should be either picking two yellow marbles or two red marbles, we have $P(YY \text{ or } RR)$

We now model the problem

$P(YY \text{ or } RR) = P(YY) + P(RR)$

$P(YY) = P(Y \text{ and } Y) = P(Y) \, P(Y/Y); P(RR) = P(R \text{ and } R) = P(R) \, P(R/R)$

Thus $P(YY \text{ or } RR) = P(YY) + P(RR) = P(Y) \, P(Y/Y) + P(R) \, P(R/R)$

Where $P(Y)$ is the probability of picking a yellow marble the first trial and picking a yellow marble in the second trial after picking a yellow marble

$P(YY \text{ or } RR) = P(Y) \, P(Y/Y) + P(R) \, P(R/R)$

In the first trial, yellow marbles are 9 and in the second trial they are 8

In the first trial, red marbles are 7 and in the second trial they are 6

$P(YY \text{ or } RR) = P(Y) \, P(Y/Y) + P(R) \, P(R/R) =$

$$\left(\frac{9}{16} \times \frac{8}{15}\right) + \left(\frac{7}{16} \times \frac{6}{15}\right) = \frac{72}{240} + \frac{42}{240} = \frac{114}{240} = \frac{19}{40} = 0.475$$

**Question 8**

A choir is composed of members who can sing soprano, alto and bass. 2 people can sing all the three voices very well. 5 people can sing both alto and bass while 6 people can sing both soprano and bass. If the choir has 28 people and 4 people who can sing bass only, what is the probability that a choir member chosen at random can sing soprano and Alto only.

**Response**

We use the venn diagram

We list what we have based on the diagram below

$K = 2$ ; $G + K = 6$,

$K + H = 5$,

$N = 4$

$M + N + P + F + H + G + K = 33$

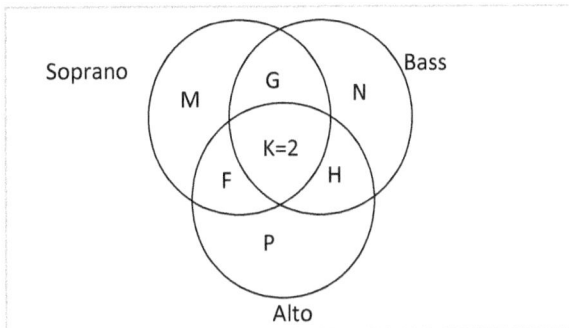

We now determine the number of people that can sing soprano and alto only

Our interest is the number $P + F + M$

From above, since $K = 2$ , $G + K = 6$, $G = 6 - 2 = 4$; $K + H = 5$, $H = 5 - 2 = 3$

Since $N = 4$, we have $G + K + H + N = 4 + 2 + 3 + 4 = 13$

$P + F + M = 28 - (G + K + H + N) = 28 - 13 = 15$

Let probability that a choir member chosen at random can sing soprano and Alto $P(S)$

Then $P(S) = \dfrac{15}{28} = 0.5357$

From the formula on multiplication rule, P(A and B) = P(A) P(B/A), we the expression for conditional

probability as $P(B / A) = \dfrac{P(A \text{ and } B)}{P(A)}$

## Question 9

In our family, 25% of us likes both pizza and candies. The percentage of our family member who like pizza is 50%, what is the probability that a family member chosen at random who likes candies also likes pizza.

## Response

We list the notations

Let the probability of those who likes pizza and candies be $P(Z \text{ and } C)$ and that of those who likes pizza be $P(Z)$

Let the probability of those whole like candies and also pizza be $P(C/Z)$

Then $P(Z \text{ and } C) = 25\% = 0.25$ ; $P(Z) = 50\% = 0.5$

Using conditional probability, we have $P(C/Z) = \dfrac{P(C \text{ and } Z)}{P(Z)} = \dfrac{0.25}{0.5} = 0.5$

The probability that a family member chosen at random who likes candies also likes pizza is 0.5

# Video Suggestions for Tutoring Sessions

Please conduct a search on either YouTube or TeacherTube to find appropriate videos for this lesson. Below are some suggested title searches:

1. Application of conditional and independent probability

2. Application of probability to explain and deal with real life activities

3. Probability is used to in decision making.

# Independent Instruction: Working on Your Own

## Questions

1. If a die is rolled once, find the probability that an outcome is not a two.

2. Two students are playing a game of die. One has to win if the sum of the outcomes is more than 10 in two rolls. What is the probability that a student wins?

3. If a die and a coin are rolled and tossed each once respectively. What is the probability that one gets a four and a head.

**1.** The number of all possible outcomes of 6

Among the six outcomes, a two can only appear once, hence

$P(\text{showing a two}) = \dfrac{1}{6}$

$P(\text{not showing a two}) = 1 - \dfrac{1}{6} = \dfrac{5}{6}$

**2.** The outcomes in any roll can be 1, 2, 3, 4, 5 and 6

The outcomes are independent hence the probability of one outcome cannot affect that of the other. The possible combinations are 5,6 ; 6,5 and 6,6. so that $5 + 6 = 6 + 5 = 11 > 10$ and $6 + 6 = 12 > 10$

5;6 means 5 in the first roll and 6 in the second roll

Probability; $P(5,6 \text{ or } 6,5 \text{ or } 6,6) = P(5,6) + P(6,5) + P(6,6) = P(5)\,P(6) + P(6)P(5) + P(6)(6)$

$= \left(\dfrac{1}{6} \times \dfrac{1}{6}\right) + \left(\dfrac{1}{6} \times \dfrac{1}{6}\right) + \left(\dfrac{1}{6} \times \dfrac{1}{6}\right) = \dfrac{1}{36} + \dfrac{1}{36} + \dfrac{1}{36} = \dfrac{3}{36} = \dfrac{1}{12}$

**3.** When a die is rolled, there are only 6 possibilities, of which one can be a four

Probability that is a four is, $P(4) = \dfrac{1}{6}$

When a coin is tossed, the probability showing a head is $P(H) = \dfrac{1}{2}$

Probability of showing a four and getting a head is $P(H \text{ and } 4) = P(H)P(4)$ since they two events are independent

$P(H \text{ and } 4) = P(H)P(4) = \dfrac{1}{2} \times \dfrac{1}{6} = \dfrac{1}{12}$

# Mini-Assessment

1.  A farmer realizes that his rectangular piece of land measuring 24 yards by 20 yards has a treasure. However, he suspects that the treasure is within a 6 yards by 9 yards rectangular piece that he identified with the help of others. What is the probability that the treasure is in the piece identified?

    **A.** 0.3409     **B.**          **C.**          **D.**          **E.** 0.1125

2.  A deck of 52 cards are shuffled and one card picks at a random, what is the probability that a card picked is not an three.

    **A.** $\dfrac{1}{13}$     **B.** $\dfrac{12}{13}$     **C.** $\dfrac{1}{52}$     **D.** $\dfrac{1}{14}$     **E.** $\dfrac{51}{52}$

3.  When a die rolled, what is the probability that the outcome is an odd number.

    **A.** $\dfrac{1}{2}$     **B.** $\dfrac{1}{6}$     **C.** $\dfrac{1}{3}$     **D.** $\dfrac{5}{6}$     **E.** $\dfrac{2}{3}$

4.  A child has 2 red marbles, 5 green and 3 yellow marbles. He picks two each at a time. What is the probability that one is a red while another is a yellow marble.

5.  A teacher has a small library where 5 books are mathematics books, 16 are history books, 4 are English book and 2 are literature books. If two books are picked at a random, what is the probability that they not history books?

**6.** In a raffle ticket, the ticket are created and numbered from 1 up to 100. The draw is then made to identify the winner. What is the probability that the winner will not have a ticket that is not a multiple of 9.

**7.** During winter season, the probability that a student will fall sick is 0.4 while the probability that a student will fail sick and fail to come to school is 0.15. What is the probability that a student will fail to come to school has fallen sick?

# Mini-Assessment Answers and Explanations

1. E

   Area of the piece of land is the sample space $= 24 \times 20 = 480 \, sq.yards$

   the area of the small identified portion is the event of interest $= 6 \times 9 = 54 \, sq.yards$

   The required probability is $= \dfrac{54}{480} = \dfrac{9}{80} = 0.1125$

2. B

   Among the 52 cards, there are 4-threes (of diamond, heart, spade and flower)

   The probability of selecting a three is $= \dfrac{4}{52} = \dfrac{1}{13}$

   The probability of not selecting a three is $= 1 - \dfrac{1}{13} = \dfrac{12}{13}$

3. A

   The outcomes from a roll of a die can be 1, 2, 3, 4, 5, 6

   Since an odd number is a number that cannot be divided by 2 completely, the numbers are 1, 3 and 5

   There are 3 possible outcomes out of 6

   The probability is $= \dfrac{3}{6} = \dfrac{1}{2}$

4. The possibilities would be $RY$ or $YR$

   The probability would be $P(RY \text{ or } YR) = P(RY) + P(YR)$

   Since the selection is without replacement, we have

   $P(RY \text{ or } YR) = P(RY) + P(YR) = P(R) \, P(Y/R) + P(Y) \, P(R/Y)$

   Probability of selecting a red for the first time is $\dfrac{2}{10}$ and that of the yellow marble for the first time is $\dfrac{3}{10}$

   The probability of selecting a yellow after selecting a red marble is $\dfrac{3}{9}$

   The probability of selecting a red after selecting a yellow marble is $\dfrac{2}{9}$

   $P(RY \text{ or } YR) = P(R) \, P(Y/R) + P(Y) \, P(R/Y) = \left( \dfrac{2}{10} \times \dfrac{3}{9} \right) + \left( \dfrac{3}{10} \times \dfrac{2}{9} \right) = \dfrac{6}{90} + \dfrac{6}{90} = \dfrac{12}{90} = \dfrac{2}{15}$

**5.** The total number of books is $16 + 5 + 4 = 25$

The number of history books is 16

The probability of selecting a history book for the first time is $\dfrac{16}{25}$

After selecting the first history book, the remain 15 and the total number of books becomes 24

The probability of selecting a history book a second time is $\dfrac{15}{24} = \dfrac{5}{8}$

The probability of selecting the two history book will be , $P(H \text{ and } H) = \dfrac{16}{25} \times \dfrac{5}{8} = \dfrac{2}{5} = 0.4$

**6.** The multiples of ten between 1 and 100 are 9, 18, 27, 36, 45, 54, 63, 72, 81, 90, 99; they are 11

The probability that a multiple of nine is selected is $\dfrac{11}{100} = 0.11$

The probability that a multiple of 9 is not selected is $1 - 0.11 = 0.89$

**7.** Let the events of falling sick and failing to come to school be $S$ and $C$

$P(S) = 0.4$

$P(S \text{ and } C) = 0.15$

We find $P(C/S)$

Using conditional probability, we have that $P(C/S) = \dfrac{P(C \text{ and } S)}{P(S)} = \dfrac{0.15}{0.4} = 0.375$

# Lesson Reflection

In this lesson, we have looked at randomness with is one of the best characteristics of occurrence of most natural events. We discussed that randomness is measured using probability with is a fraction. We looked at various probability models and rules used to simply probability problems. One of the special rules were the conditional probability that we discussed towards the end of the lesson. Finally we have looked at the applications of probability to real life.

# Lesson 5
# Random Variables, Binomial & Geometric Distributions

## Lesson Description

This lesson is designed to help students analyze different types of data that includes normal distributions. Please be sure to utilize the questions to help spark student engagement and cover the vocabulary that is associated with this specific tutoring session. For your own knowledge, sample responses have been provided to guide you as well.

## Learning Objectives

In today's session, the learner will analyze different types of data that includes normal distributions with at least 75% or above accuracy in 3 out of 4 trials.

## Connect Learning Objectives to Students' Lives

**A.** To familiarize and differentiate between discrete and continuous random variables.

**B.** To familiarize with the formulas of discrete and continuous random variables.

**C.** To discuss the concept of large numbers.

**D.** To determine the characteristics of random variables in a binomial and geometric settings.

**E.** To highlight some applications of random variables, laws of large numbers, binomial and geometric setting.

# Introduction

An experiment, which in our lesson, we will refer to statistical experiment, tries to study a number of characteristics of variables. These may be mean, mode, range, variance among others. In perfect statistical experiment, these variable are not assigned values but take values based on the chance of their occurrence. Therefore, the assignment of these values to these variables is termed as random. Such variables described above are called random variable. We would like to find out some distributions and features of these variables when tuned to some specific conditions, such as a geometric and binomial conditions.

# Specific Vocabulary for Tutoring Session

### Discrete Random Variable

It is a random variable that take values from a countable set.

### Continuous Random Variable

It is a random variable that takes any value within a given range of values.

# Direct Instruction: Modeling For You

### Discrete and Continuous Variables

Random variables can be classified into two categories, the discrete and continuous random variable. Let us consider a set of countable values, for instance, the number of students, the number of families, the number of people having red white, the number of malaria infections among others. When a random variable takes values from such countable set, we say it is a **discrete random variable**. Discrete variables, in most cases, takes integers values.

On the other hand, when the random variable takes any value between the given range, we call it a **continuous random variable**. This implies that the value may be a decimal number. For instance, if a variable represents the height of a person which falls between 50 inches and 100 inches, the variable is a continuous random variable.

## Question 1

Can a continuous random variable take an integer value?

### Response

Yes. When the range of values on which the continuous random variable is defined contains an integer or integers, then the variable can take an integers value. Therefore, it all depends on the range of values the variable is defined. For instance, when the random variable $X$ is the mass of an object between 2.1 pounds and 7.5 pounds, the variable is a continuous variable which can take the integer values such as 3, 4, 5, 6 and 7. If the range of values were 3.12 pounds and 3.95 pounds, then the variable would not take an integer value.

## Question 2

Classify the following situations as having discrete or continuous random variable.

**a).** The outcome of rolling a fair die

**b).** The height of an adult person

**c).** Selecting a given number of students from a class

**d).** Time taken by an athlete to complete a race

### Response

Random variables can be classified into two categories, the discrete and continuous random variable. Let us consider a set of countable values, for instance, the number of students, the number of families, the number of people having red white, the number of malaria infections among others. When a random variable takes values from such countable set, we say it is a **discrete random variable**. Discrete variables, in most cases, takes integers values.

On the other hand, when the random variable takes any value between the given range, we call it a **continuous random variable**. This implies that the value may be a decimal number. For instance, If a variable represents the height of a person which falls between 50 inches and 100 inches, the variable is a continuous random variable.

## Probability Distributions

The random variables, whether discrete or continuous, are associated with specific probabilities in a statistical experiments. When we have a table giving the probability of different outcomes of variables, we call it a probability distribution. This is usually the case with discrete variables. For the case of continuous random variables, an equation relating the two, the outcomes and the probability is given.

The sum of probabilities in a probability distribution is 1.

We will use capital letter $X$ for the variable and the small letter $x$ for the specific values the variable $X$ takes. Thus $P(X = x)$ is read as the probability that the random variable $X$ is $x$ where $x$ can be 0, 1,…. or 6 as in the following example.

Consider the following illustration

Let $X$ be a random variable which takes values 0, 1, 2, 3, 4, 5 or 6 with probabilities 0.1, 0.08, 0.24, 0.2, 0.26, 0.12.

The table of the distribution will be

| Outcomes, $X$ | 0 | 1 | 2 | 3 | 4 | 5 | 6 |
|---|---|---|---|---|---|---|---|
| Probabilities, $P(X = x)$ | 0.1 | 0.08 | 0.15 | 0.2 | 0.26 | 0.12 | 0.09 |

The distribution is a probability distribution since the sum of the probabilities is 1.

We would like to discuss the concepts that $P(X = x)$, $P(X > x)$, $P(X < x)$, $P(X \leq x)$, $P(X \geq x)$

We will take a fixed point in all the above cases. Let $x = 4$.

$P(X = 4)$

This the probability that $X = 4$. This can be read directly from the table. $P(X = 4) = 0.26$

$P(X < 4)$

From the table above, $X < 4$ implies $X = 0, 1, 2,$ and 3 not including 4.

Therefore, $P(X < 4) = P(X = 0) + P(X = 1) + P(X = 2) + P(X = 3) = 0.1 + 0.08 + 0.15 + 0.2 = 0.53$

Also, $P(X < 4) = 1 - (P(x = 4) + P(x = 5) + P(x = 6)) = 1 - (0.26 + 0.12 + 0.09) = 1 - 0.47 = 0.53$

$P(X > 4)$

From the table above, $X > 4$ implies $X = 5,$ and 6 not including 4.

Therefore, $P(X > 4) = P(X = 5) + P(X = 6) = 0.12 + 0.09 = 0.21$

Also, A $P(X > 4) = 1 - (P(x = 4) + P(x = 3) + P(x = 2) + P(x = 1) + P(x = 0)) = 1 - 0.79 = 0.21$

$P(X \leq 4)$

From the table below, $X \leq 4$ implies $X = 0, 1, 2, 3$ and 4.

But $P(X \leq 4) = 1 - (P(X = 5) + P(X = 6)) = 1 - (0.12 + 0.09) = 0.79$

$P(X \geq 4)$

From the table below, $X \geq 4$ implies $X = 4$, 5, and 6.

$P(X \geq 4) = P(X = 4) + P(X = 5) + P(X = 6) = 0.26 + 0.12 + 0.09 = 0.47$

## Mean and Variance of Random Variables

The mean of a random variable is termed as the expected value and denoted E(X) (sometimes denoted $\mu$ (mu)), read as the **expected value of X**. It implies the average of the random variable.

By considering the above illustration, we have 7 outcomes of the random variable. We can therefore index the random variables as $X_1 = 0, X_2 = 1, ..., X_7 = 6$,

The respective probabilities will be $P(X_1) = 0.1, P(X_2) = 0.08, ..., P(X_7) = 0.09$,

We can take $X_1$ and their respective probabilities to be $P(X_1)$, where $i = 0, 1, 2, ...6$

The expected value of the random variable X will be $E(X_i) = \sum_{i=1}^{n} X_i P(X_i)$

The variance of a random variable X is $V(X) = E(X - \mu)^2 = E(X^2) - \mu^2$

## Properties of the Expected Value

**(i).** The expected value of the sum of two independent random variables $X$ and $Y$ is

$E(X + Y) = E(X) + E(Y)$

**(ii).** The expected value of a multiple of a random variables $X$ is $E(bY) = bE(Y)$

Where $b$ is a constant number

## Question 3

The table below shows the distribution of a random variable Y. Determine the expected value of the variable and the expected value of twice the random variable.

| Y | 0 | 1 | 2 | 3 |
|---|---|---|---|---|
| P(Y = y) | 0.22 | 0.27 | 0.23 | 0.28 |

**Response**

We determine $E(X)$ and $E(2X)$

$$E(X_i) = \sum_{i=1}^{n} X_i P(X_i) = (0 \times 0.22) + (1 \times 0.27) + (2 \times 0.23) + (3 \times 0.28) = 1.57$$

$$E(2X) = 2\,E(X) = 2 \times 1.57 = 3.14$$

**(iii).** When two random variables are independent, the expected value of their product would be the product of their expected values.

$$E(XY) = E(X)\,E(Y)$$

## Properties of Variance

**(i).** The variance of the sum of two independent random variables $X$ and $Y$ is

$$V(X + Y) = V(X) + V(Y)$$

**(ii).** The expected value of a multiple of a random variables $Y$ is

$$V(bY) = b^2\,V(Y)$$

Where $b$ is a constant number

**(iii).** When a constant number is added to a random variable, then the variance of the sum is equal to the variance of the variable.

$$V(b + Y) = V(Y)$$

Where $b$ is a constant number

## Questions 4

A company manufactures and pack toothpaste in tubes. The weight of the whole package varies based on the temperature at which it was packed. The following table shows the temperature at different cases and the probability that the package will have the required mass at a given temperature. Determine the variance of the distribution.

| Temperature (°C) | 43 | 44 | 45 | 46 | 47 | 48 |
|---|---|---|---|---|---|---|
| Probability | 0.12 | 0.18 | 0.29 | 0.17 | 0.14 | 0.1 |

We have $V(X) = E(X^2) - \mu^2$ where

$$= (43 \times 0.12) + (44 \times 0.18) + (45 \times 0.29) + (46 \times 0.17) + (47 \times 0.14) + (48 \times 0.1)$$

$\mu = E(X) = 45.33°C$

$\mu^2 = 45.33^2 = 2054.81$

$$E(X^2) = (43^2 \times 0.12) + (44^2 \times 0.18) + (45^2 \times 0.29) + (46^2 \times 0.17) + (47^2 \times 0.14) + (48^2 \times 0.1)$$

$$= 2056.99$$

$$V(X) = E(X^2) - \mu^2 = 2056.99 - 2054.81 = 2.18$$

## Laws of Large Numbers

**The laws of large numbers are relations that predicts the expected value, μ, of a probability** experiment that is continuously done in many trials. The number of trials done is what is referred to and has to be a large number. There are two laws of large numbers, namely, the weal and the strong law of large numbers.

## The Weak Law of Large Numbers

The weak low allows a small margin within which the sample mean and the expected value of the random variable should be. The probability that the difference between the two values lies in the margin is very high thus compatible to 1. On the other hand the probability that the difference between the two values lies outside the margin is zero.

It states as follows

Let $\overline{X}_n$ be the sample mean of a random variable $X$ and $\mu$ the expected value of $X$, $(E(X) = \mu)$.

$\overline{X}_n$ approaches μ as n approaches infinity

For sufficiently small positive value ε the limit

$$\lim_{x \to \infty} P(|\overline{X}_n - \mu| \leq \varepsilon) = 1 \text{ equivalently } \lim_{x \to \infty} P(|\overline{X}_n - \mu| > \varepsilon) = 0$$

## The Strong Law of Large Numbers

The strong law of large numbers gives a surety that the sample mean will converge to expected value of after a sufficiently many probability experiments are done.

It states as follows

Let $\overline{X}_n$ be the sample mean of a random variable $X$ and $\mu$ the expected value of $X$, $(E(X) = \mu)$.

Then

$$P(\lim_{x\to\infty} \overline{X}_n = \mu) = 1 \quad \text{equivalently} \quad P(\lim_{x\to\infty} |\overline{X}_n - \mu|) = 1$$

# Direct Instruction: Working With You

## Question 5

State the difference between the weak and the strong law of large numbers

## Response

In the weak law of large numbers, there is a margin between the expected value of the random variable and the limit of sample mean while for the strong law of large numbers, it does not exists. This implies that for the former, all the probabilities for each sample are concentrated at the expected value while for the later, they fall within margin $\mu \pm \varepsilon$.

## Binomial Setting

A binomial setting is a probability situation where there are two outcomes in every trial where

**(i).** The probability of a particular outcome (success) remains the same in all trials. The probability of success is equal in all the trials.

**(ii).** The trials are independent

**(iii).** The number of trials, n, is predetermined before the process, hence it is fixed.

The interest is usually to determine the probability of the number of successes recorded in the n independent trials conducted

When a process satisfy the above conditions, it is said to have a **binomial setting**.

## Questions 6

Explains why or why not choosing a cream tennis ball from a container having cream and yellow tennis ball is a binomial setting.

## Response

We determine if the outcomes are binary

When we consider the first trial, we either choose a cream tennis ball or a ball that is not cream. Hence we have two cases, a binary case.

We then determine if the trials are independent

After choosing the first ball, the number of balls reduces hence the probability of choosing the next ball will not be the same as the first. Therefore, the trials are not independent.

This is enough to prove that the case is not a binomial setting.

## Binomial Probability

In the discussion above, the number of success in the n trials is the binomial random variable. The binomial distribution refers to the distribution of the binomial random variable whose parameters are the number of independent trial made and the probability of success in one outcome.

Since $X$, the binomial random variable, is the number of successes, we can only have X taking a maximum of N and a minimum of 0.

Having the above information, we now define the binomial probability.

Let $n$ be the number of independent trials and $X$ the binomial random variable where $X$ can be 0, 1, 2, … or $n$. Let $p$ the probability of success and $1 - p = q$ the probability of failure, if $k$ is the number of successes in n independent trials, the probability of $k$ success is given by

$$P(X = k) = \left( \frac{n!}{k!(n-k)} \right) p^k q^{n-k}$$

The expected value and the variance of a binomial random variable are given by

$$E(X) = np$$

$$V(X) = \sqrt{npq}$$

## Question 7

$Y$ is a binomial variable where $n = 5$ and $p = 0.6$. Find the mean and the variance of the variable.

## Response

We list what we have and use it to substitute in the formulas for mean and variance

$n = 5$ and $p = 0.6$

The mean $= E(X) = np = 5 \times 0.6 = 3.0$

$p = 0.4$ hence $q = 1 - p = 1 - 0.6 = 0.4$

The variance $= \sqrt{npq} = \sqrt{(5 \times 0.6 \times 0.4)} = \sqrt{1.2} = 1.095$

## Question 8

A fair coin is tossed seven times. Find the probability that the outcomes had more than 5 tails.

## Response

We determine the independence of the trials

The outcome, tail is taken as a success

The outcome in any trial is not affected by that in another trial, hence the trails are independent

We list what we are given in the question

The probability of success in one trial is p = 0.5

The probability of failure $= q = 1 - p = 1 - 0.5 = 0.5$

The number of independent trials $= n = 7$

The number of success $= k > 5$

We formulate the problem

$P$(More than 5 tails) $= P(X > 5) = P(X = 6) + P(X = 7)$

We use the information given to solve the problem

$$P(X = 6) = \left( \frac{7!}{6!(7-6)} \right) 0.5^6 \times 0.5^{7-6} = 0.055$$

$$P(X = 7) = \left( \frac{7!}{7!(7-7)} \right) 0.5^7 \times 0.5^{7-7} = 0.0078$$

$P(X > 5) = P(X = 6) + P(X = 7) = 0.055 + 0.0078 = 0.0628$

## Geometric Setting

A binomial setting is a probability situation where there are two outcomes in every trial such

**(i).** The probability of a particular outcome (success) remains the same in all trials. The probability of success is equal in all the trials.

**(ii).** The trials are independent

The interest is usually to determine the probability of the number of trials required to get the first success. This number of trials is called the **geometric random variable**.

When a process satisfy the above conditions, it is said to have a **geometric setting**.

## Geometric Probability Distribution

This is the distribution of a geometric random variable whose parameter is the probability of success. If $p$ and $q$ are the probabilities of success and failure respectively and $k$ is the number of trials made before the first success, then $P(X = k) = pq^{k-1}$

$$\text{where } q = 1 - p$$

The mean and variance of geometric distribution is $E(X) = \dfrac{1}{p}$

---

### Question 9

A fair six sided die is rolled till a four shows up. What is the probability that a four shows up in the fifth roll?

---

### Response

We determine if it is a geometric problem and list what we have

When a die is rolled, it may or it may not show a four, hence it is a binary case.

The outcomes in any trial will not affect that of the other, hence the trials are independent

Let the outcome of a 4 represents a success, then we determine the probability when the first success first comes hence the case has a geometric setting. We apply the geometric formula.

For a six sided die, the probability of showing a four is $p = \dfrac{1}{6}$

The number of times the die is rolled is $k = 5$

The probability of failure $= 1 - p = \dfrac{5}{6}$

We use the information to determine the probability

$$P(X = 5) = \frac{1}{6} \times \left(\frac{1}{6}\right)^{5-1} = \frac{1}{6} \times \left(\frac{1}{6}\right)^{4} = \frac{1}{7776} = 0.0001286$$

# Video Suggestions for Your Tutoring Session

Please conduct a search on either YouTube or TeacherTube to find appropriate videos for this lesson. Below are some suggested title searches:

1. Use of random variable to determine the fairness and the viability of the game

2. The use of laws of large numbers to determine the future performance of the business

3. Using binomial probability in gambling

# Independent Instruction: Working On Your Own

## Questions

1. Identify if the following variable is discrete or continuous.

   $X$ is a random variable showing the number of middle class people in a suburb.

2. The variance of a random variable is 2.5. If the expectation of the square of the random variable is 227.5, find the expectation of the variable.

3. The distribution of a random variable $X$ is as shown in the following figure. What will be the expectation of $5X$.

| $X$ | 0 | 1 | 2 | 3 |
|---|---|---|---|---|
| $P(X = x)$ | 0.14 | 0. 36 | 0.28 | 0.22 |

**1.** We find out if $X$ can be a counting number or not

The number of middle class people in a given area can be counted hence $X$ takes an integer value.

Therefore, $X$ is a discrete random variable

**2.** We first list what we have

$V(X) = 2.5$, $E(X^2) = 227.5$

We the apply the formula for variance to determine the expected value

$X) = 2.5$ , $E(X^2) = 227.5$

$V(X) = E(X^2) - \mu^2$

Upon substitution, we have

$2.5 = 227.5 - \mu^2$

Hence $\mu^2 = 227.5 - 2.5 = 225$

$\mu = \sqrt{225} = 15$

**3.** We apply the formulas $E(X_i) = \sum_{i=1}^{n} X_i\, P(X_i)$ and $E(bX) = b\, E(X)$ where $b$ is a constant

$E(X_i) = \sum_{i=1}^{n} X_i\, P(X_i) = (0 \times 0.14) + (1 \times 0.36) + (2 \times 0.28) + (3 \times 0.22) = 1.58\,]$

$E(bX) = b\, E(X) = 5 \times 1.58 = 7.9$

# Mini-Assessment

1.  A random variable $X$ is such that the expectation of its square is 3382 and its expected value is 18. Determine the variance of $\dfrac{X}{3}$.

    **A.** 18         **B.** 2         **C.** 6

    **D.** 3         **E.** 1

2.  The temperature of the body of a normal human being is a random variable with the following distribution. What is the expected temperature of a normal human being based on the following data?

| Temperature, T in °C | 36. 5 | 36.7 | 36.9 | 37.1 | 37.3 |
|---|---|---|---|---|---|
| $P(T = t)$ | 0.1 | 0. 22 | 0.26 | 0.30 | 0.12 |

    **A.** 37.10         **B.** 37.05         **C.** 36.97

    **D.** 37. 05         **E.** 36.92

3   A teacher has arranged grouped a class based on the performance in the test. He wishes to choose a person from the group to make a representation, what is the probability that the chosen person scored more than 59% if the score were rounded off to the nearest whole number?

| Sores, S, in percentage | < 50 | 50 - 59 | 60 – 69 | 70 – 79 | > 80 |
|---|---|---|---|---|---|
| $P(S = S)$ | 0.04 | 0. 18 | 0. 20 | 0.34 | 0. 24 |

    **A.** 0.22         **B.** 0.18         **C.** 0.78

    **D.** 0.96         **E.** 0.20

4. Identify the kind of random variable in the following statement,

   The random variable is a point in the rectangular plot whose length is 2 mile and width is 1.5 mile.

5. Find the probability that in three tosses, there will be two heads.

6. A storage facility has six sections. A farmer would like to locate which store has chicken feed. He does this by searching in all the stores. If we are interested in the number of stores he searches in before getting the store in the feed are stored, find out if the probability setting is a binomial, geometric or none of them.

7. A fair die is rolled three times. Find the expected value of the random variable that a 5 shows up in the third roll.

# Mini-Assessment Answers and Explanations

**1.** B

$V(X) = E(X^2) - \mu^2$ , but $E(X^2) = 3382$ and $\mu = 18$

$V(X) = 3382 - 58^2 = 18$

$V(cX) = c^2 V(X)$

$V(\frac{1}{3}X) = \frac{1}{9}V(X) = \frac{1}{9} \times 18 = 2$

**2.** E

**3.** C

$P(\text{More than } 69) = 1 - P(S \le 59) = 1 - P(S < 50) - P(50 \le S \le 59)$

$$= 1 - 0.04 - 0.18 = 0.78$$

**4.** Continuous random variable

There are numerous points in the range given. Since we can pick any point within the area, the variable is a continuous random variable.

5. 0.1875

Let the number of heads be the success. The probability of success in each trial is $p = 0.5$

Since the outcome in one trial does not affect the other, the trials are independent.

The probability of failure $q = 1 - p = 1 - 0.5 = 0.5$

The number or rolls, $n = 3$.

The number of success is $k = 2$.

The problem is a binomial probability problem.

We use the formula

$P(X = k) = \left( \dfrac{n!}{k!(n-k)} \right) p^k q^{n-k}$

$P(X = 2) = \left( \dfrac{3!}{2!(3-2)} \right) 0.5^3 \times 0.5^{3-2} = 3 \times 0.5^4 = 0.1875$

**6.**   The setting is neither binomial nor geometric.

The success implies getting the location of the feed in a given search.

The probability of the success in the first search $= \dfrac{1}{6}$

After one search, he remains with 5 sections to check.

The probability of success in the next search given that he does not get in the first search is $\dfrac{1}{5}$

Since the probability of success in all trials is not equal, the trials are dependent therefore, the setting is neither binomial nor geometric.

**7.**   6

Let the success be when a 5 show up

The probability of success $p = \dfrac{1}{6}$

The random variable $X$ is the number of rolls before up to the first success. Thus X = 3.

This is a geometric distribution whose expected value is $E(X) = \dfrac{1}{p} = 6$

# Lesson Reflection

In today's lesson, we have learned about random variables and categorized them into discrete and continuous random variables. We went further to discuss their mean and variance. This was followed by the laws that describes the expected value of a random variable after a large number of samples are investigated. The laws are called the laws of large numbers. We have finalized the lesson by looking at the binomial and geometric setting and the applications of the lesson in day to day life.

www.ingramcontent.com/pod-product-compliance
Lightning Source LLC
Chambersburg PA
CBHW051415200326
41520CB00023B/7241